Applied Geospatial Data Science with Python

Leverage geospatial data analysis and modeling to find unique solutions to environmental problems

David S. Jordan

BIRMINGHAM—MUMBAI

Applied Geospatial Data Science with Python

Publishing Product Manager: Dinesh Chaudhary
Content Development Editor: Shreya Moharir
Technical Editor: Devanshi Ayare
Copy Editor: Safis Editing
Project Coordinator: Farheen Fathima
Proofreader: Safis Editing
Indexer: Pratik Shirodkar
Production Designer: Ponraj Dhandapani
Marketing Coordinators: Shifa Ansari, Vinishka Kalra

First published: February 2023

Production reference: 1270123

Published by Packt Publishing Ltd.
Livery Place
35 Livery Street
Birmingham
B3 2PB, UK.

ISBN 978-1-80323-812-8

www.packtpub.com

To those who came before me and those who come after, leaving an indelible mark on the world and making it a better place.

Acknowledgments

I'd like to acknowledge the many teachers, professors, and mentors who have shared their infinite knowledge and wisdom with me throughout the years. Without them, my path through life would likely be completely different. I'd like to start with Dr. Jennifer Stuart, whose marketing analytics course introduced me to the power of data science for the first time. I'd also like to thank Dr. Connie Rothwell and Delbridge Narron for supporting me in my honors thesis and encouraging me to question and challenge the world around me. I'd also like to thank the staff and professors at the Institute for Advanced Analytics at North Carolina State University for providing me with a first-class applied data science education.

I'd be remiss if I didn't thank my friends and family for supporting me in this endeavor, as well as all of the other crazy leaps I've taken throughout my personal and professional life – most importantly, my older brother, Jeff Jordan, who has encouraged me in my successes and failures and listened to me vent in my times of need. Thank you for also continuing to be the intellectual springboard that has led to some of my best thinking.

Lastly, I'd like to thank the many developers who are actively and often thanklessly building out the open source spatial data science ecosystem. Without these developers, this book would not have been possible, as we wouldn't have had the tools or data at our disposal. I encourage you, the reader of this book and bourgeoning spatial data scientist, to give back to the community in any way possible in the future.

Contributors

About the author

David S. Jordan has made a career out of applying spatial thinking to tough problem spaces in the domains of real estate planning, disaster response, social equity, and climate change. He currently leads distribution and geospatial data science at JPMorgan Chase & Co. In addition to leading and building out geospatial data science teams, David is a patented inventor of new geospatial analytics processes, a winner of a **Special Achievement in GIS (SAG)** Award from Esri, and a conference speaker on topics including banking deserts and how great businesses leverage GIS.

About the reviewer

Rohit Singh has been working in the field of geospatial analysis and modeling for the past 5 years, and he is currently working as a geospatial data scientist at Near Intelligence Pvt. Ltd. In his current role, he is in charge of all spatial data features and engineering pipelines, as well as curating spatial data from different sources, developing methods for spatial data processing in an optimized manner. He has developed methods in Apache Spark to handle the processing and modeling of spatial big data, and he regularly contributes to the GIS and data science communities. He developed a Python module to optimize the geohash generation process for polygons. In the future, he wants to contribute more to spatial data science by developing more spatial data methods and models.

Table of Contents

Part 3: Geospatial Modeling Case Studies

8

Spatial Clustering and Regionalization   165

Preface

By the time this book has been published, the world will have just formally exited a global pandemic, and society as a whole will be trying to grapple with the new normal in the post-COVID era. During the depths of the pandemic, spatial analysis was featured in prime time through the great work of **Johns Hopkins University** (**JHU**)'s COVID-19 dashboard, which can be found at `https://coronavirus.jhu.edu/map.html`. The JHU dashboard monitored the spread of the virus across the globe in near real time, and this map was likely the first time that the masses were exposed to the power of spatial analysis, spatial data visualization, and spatial data science. However, spatial analysis has long been used to analyze the spread of diseases. In fact, way back in 1854, John Snow produced a map of cholera deaths in London, which allowed him to show that cholera was spread through germs in water wells and not through miasma in the air, as many thought during that time.

Reeling from this global pandemic is not the only problem that our modern society faces. Today, supply chain issues that face economies across the globe are driving inflation to heights not seen in several decades. In addition to this, climate change is causing major rivers across the globe to dry up, including the Colorado and Mississippi rivers in the United States, the Yangtze in China, the Rhine in Germany, and the Danube in Romania. Climate change is also leading to more extreme weather events, yielding devastating flooding in areas such as Florida in the United States and Pakistan in South Asia.

We are also living through a time in which more and more people are willing to stand up for equity and call out inequities when they see them. In the United States and across the world, teams of people are researching high-profile inequities in terms of the global food supply, healthcare access, and financial services. Others are looking into lesser-known inequities, such as urban heat islands and lack of shade. Collectively, teams of this kind are working hard to ensure that future generations won't face the inequities of their forefathers.

We now have the data, tools, and technology to begin to do something about each of these problems. Spatial analysis and data science have the potential to provide enormous value in helping us find solutions, perform resiliency planning, and better educate ourselves and those around us. However, while performing spatial analysis and producing compelling visualizations is now easier than ever, it is not without risks. By nature, maps and spatial data are representations of real-world processes and are often incomplete or can easily be manipulated and thus the truth can be distorted. One recent example of map manipulation happened in an event that has since been dubbed "Sharpie-gate," in which then-President Donald Trump altered an NOAA hurricane path map with a Sharpie in defiance of the scientific community. While this example may seem comical, there are many nuances to spatial analysis, data science, and cartography that you'll need to be aware of as a burgeoning spatial data scientist.

This book is written for data scientists seeking to incorporate geospatial analysis into their work and for **geographic information system** (**GIS**) professionals seeking to incorporate data science methods into their work. Our goal is that this text will help these communities to develop a common understanding and shared vernacular, enabling them to properly incorporate geographic context into modeling, analysis, and visualization.

This book will begin with the fundamentals of GIS and data science before moving into detailed examples of spatial data science workflows built upon practical applications of geospatial data science that are industry agnostic. We will begin by teaching you the fundamentals of sourcing and working with geospatial data. Building upon this, we will teach you how to integrate spatial data and spatial thinking into your data science processes to hopefully improve model performance and develop a more accurate representation of the world around us.

We hope that you, as a member of the next generation of spatial data scientists, are empowered to leverage spatial thinking and analysis, which may help us find solutions to the problems currently facing our society and better prepare for the future ahead.

Who this book is for

This book is for you if you are a data scientist seeking to incorporate geospatial thinking into your workflows or a GIS professional seeking to incorporate data science methods into yours. You'll need to have a foundational knowledge of Python for data analysis and/or data science.

What this book covers

Chapter 1, Introducing Geographic Information Systems and Geospatial Data Science, lays the foundations for the book by introducing you to GIS and its commonalities with and differences from geospatial data science. In this chapter, we also walk through the data science pipeline that you'll follow throughout the book.

Chapter 2, What Is Geospatial Data and Where Can I Find It?, introduces you to common geospatial data types and formats that you'll work with throughout your geospatial data science workflows. In this chapter, we'll also introduce various categories of geospatial data, ranging from human geography to country- and area-specific data.

Chapter 3, Working with Geographic and Projected Coordinate Systems, will introduce you to geographic and projected coordinate systems and help you avoid some of the most common pitfalls of working with geospatial data.

Chapter 4, Exploring Geospatial Data Science Packages, covers a wide variety of Python geospatial data science packages that allow you to perform spatial data processing, analysis, visualization, and modeling.

Chapter 5, Exploratory Data Visualization, shows you how to harness the power of spatial data to create compelling static and dynamic mapping applications.

Chapter 6, Hypothesis Testing and Spatial Randomness, introduces you to the topic of complete spatial randomness and a variety of statistical tests to better understand whether your data reflects patterns across space.

Chapter 7, Spatial Feature Engineering, will walk you through how to derive new spatial-based features known as summary spatial features and proximity spatial features from both tabular and geo-enabled data assets.

Chapter 8, Spatial Clustering and Regionalization, introduces you to a class of unsupervised machine learning models known as clustering models, through which you'll create spatial clusters and regions from your data.

Chapter 9, Developing Spatial Regression Models, will open your eyes to the power that spatial data can bring to regression models through the incorporation of spatial effects.

Chapter 10, Developing Solutions to Spatial Optimization Problems, will show you how to use linear programming in combination with spatial data to solve problems such as the Vehicle Routing Problem and the Location Set Covering Problem.

Chapter 11, Advanced Topics in Spatial Data Science, covers more advanced topics in spatial feature engineering, spatial modeling, and spatial ethics.

To get the most out of this book

As readers of this book, we assume that you come from a background in either data science or GIS. We also expect that you have some foundational knowledge of working with Python.

Software/hardware covered in the book	Operating system requirements
Anaconda Distribution	Windows, macOS, or Linux
Python 3.10.6	Windows, macOS, or Linux

Additionally, you will need to set up keys to several APIs, from which you will access data throughout the book.

API	Setup link
OpenMapQuest	`https://developer.mapquest.com/user/login/sign-up`
Google Maps	`https://developers.google.com/maps`
US Census Bureau	`https://api.census.gov/data/key_signup.html`

If you are using the digital version of this book, we advise you to type the code yourself or access the code from the book's GitHub repository (a link is available in the next section). Doing so will help you avoid any potential errors related to the copying and pasting of code.

The quality of the hardware can impact the runtime for some analyses, as is the case for most data science activities. As such, we recommend hardware similar or better to the specified hardware outlined to prevent any potential issues:

- NVIDIA GeForce GTX 1050

- 16 GB RAM

We recommend that you use Anaconda as your Python environment and package manager. To begin installing the Anaconda Distribution, you'll want to visit the Anaconda Distribution installation website at `https://docs.anaconda.com/anaconda/install/`. The Python version we are using throughout this book is 3.10.6, as this is one of the latest versions of Python available at the time of publication. Leveraging this version will ensure that all packages are compatible. To make the setup of your virtual environment as streamlined as possible, we've exported our `environment.yml` file and uploaded it to the GitHub repository at `https://github.com/PacktPublishing/Applied-Geospatial-Data-Science-with-Python`.

To set up the virtual environment called `GeospatialPython`, launch Anaconda prompt and execute the following command:

```
conda env create -file environment.yml
```

You'll need to substitute `environment.yml` for the full path of the downloaded file.

After the environment is installed, you can activate it by executing the following command:

```
conda activate GeospatialPython
```

Throughout the book, you'll see the following code:

```
data_path = r'YOUR FILE PATH'
```

Anytime you see this, you'll need to substitute 'YOUR FILE PATH' with the file path of the data folder which can be downloaded from the GitHub repo. The data stored in the GitHub repo can be found in the **Releases** section or by visiting `https://github.com/PacktPublishing/Applied-Geospatial-Data-Science-with-Python/releases`. There are three parts to the data:

- `Data.pt1.zip`

- `LCMS_CONUS_v2021-7_Land_Cover_Annual_2021.zip`

- `S2B_MSIL2A_20220504T161829_N0400_R040_T17TNF_20220504T210702.SAFE.zip`

You'll need to extract the contents of these zip folders and store the contents in a single folder. You'll then point to this folder any time you see 'YOUR FILE PATH' referenced in the Jupyter notebooks.

Similarly, you will also see the following code from time to time:

```
out_path = r"YOUR FILE PATH"
```

You'll need to substitute YOUR FILE PATH in this code reference with the directory to which you'd like the output to be saved.

Download the example code files

You can download the example code files for this book from GitHub at https://github.com/PacktPublishing/Applied-Geospatial-Data-Science-with-Python. If there's an update to the code, it will be updated in the GitHub repository.

We also have other code bundles from our rich catalog of books and videos available at https://github.com/PacktPublishing/. Check them out!

Download the color images

We also provide a PDF file that has color images of the screenshots and diagrams used in this book. You can download it here: https://packt.link/AN9bG.

Conventions used

There are a number of text conventions used throughout this book.

Code in text: Indicates code words in text, database table names, folder names, filenames, file extensions, pathnames, dummy URLs, user input, and Twitter handles. Here is an example: "The PyProj package is useful when working with cartographic projections and geodetic transformations."

A block of code is set as follows:

```
world_ae = world.to_crs("ESRI:54032")
graticules_ae = grat.to_crs("ESRI:54032")
```

When we wish to draw your attention to a particular part of a code block, the relevant lines or items are set in bold:

```
patients = 150 # number of demand points represented as
patients
medical_centers = 4 # number of service points represented as
medical centers
```

Bold: Indicates a new term, an important word, or words that you see onscreen. For instance, words in menus or dialog boxes appear in **bold**. Here is an example: "We've called ours **VRP Project**. Select this project and scroll down to **Enabled APIs**. Then, select **Directions API** and click **Enable**. "

> **Tips or important notes**
> Appear like this.

Get in touch

Feedback from our readers is always welcome.

General feedback: If you have questions about any aspect of this book, email us at `customercare@packtpub.com` and mention the book title in the subject of your message.

Errata: Although we have taken every care to ensure the accuracy of our content, mistakes do happen. If you have found a mistake in this book, we would be grateful if you would report this to us. Please visit `www.packtpub.com/support/errata` and fill in the form.

Piracy: If you come across any illegal copies of our works in any form on the internet, we would be grateful if you would provide us with the location address or website name. Please contact us at `copyright@packt.com` with a link to the material.

If you are interested in becoming an author: If there is a topic that you have expertise in and you are interested in either writing or contributing to a book, please visit `authors.packtpub.com`.

Share Your Thoughts

Once you've read *Applied Geospatial Data Science with Python*, we'd love to hear your thoughts! Scan the QR code below to go straight to the Amazon review page for this book and share your feedback.

https://packt.link/r/1-803-23812-7

Your review is important to us and the tech community and will help us make sure we're delivering excellent quality content.

Download a free PDF copy of this book

Thanks for purchasing this book!

Do you like to read on the go but are unable to carry your print books everywhere? Is your eBook purchase not compatible with the device of your choice?

Don't worry, now with every Packt book you get a DRM-free PDF version of that book at no cost.

Read anywhere, any place, on any device. Search, copy, and paste code from your favorite technical books directly into your application.

The perks don't stop there, you can get exclusive access to discounts, newsletters, and great free content in your inbox daily

Follow these simple steps to get the benefits:

1. Scan the QR code or visit the link below

https://packt.link/free-ebook/9781803238128

2. Submit your proof of purchase
3. That's it! We'll send your free PDF and other benefits to your email directly

Part 1:
The Essentials of
Geospatial Data Science

To begin your journey as a geospatial data scientist, it is critical that you familiarize yourself with the critical elements of geospatial data science. In this part of the book, we'll define geospatial information systems and their relationship to data science and geospatial data science. You'll also be introduced to various types of geospatial data and a plethora of open geospatial data resources. Later in this part, you'll be introduced to geographic and projected coordinate systems, which are how we model the earth in different mediums. We'll conclude the chapter with an introduction to the Python packages you'll leverage throughout Part 2 and Part 3 of the book.

This part comprises the following chapters:

- *Chapter 1, Introducing Geographic Information Systems and Geospatial Data Science*

- *Chapter 2, What Is Geospatial Data and Where Can I Find It?*

- *Chapter 3, Working with Geographic and Projected Coordinate Systems*

- *Chapter 4, Exploring Geospatial Data Science Packages*

1

Introducing Geographic Information Systems and Geospatial Data Science

It is estimated that human society generates several quintillion bytes of data every day. The amount and speed with which our society generates data are also estimated to increase yearly as more and more devices become connected. Devices in the palms of our hands generate rich data assets ranging from detailed human movement data to data on purchasing behavior that connects online transactions with those made at physical storefronts. At the same time, **remote sensing** devices located outside our atmosphere are generating detailed images, known as **satellite imagery**, of the Earth at a 0.5-meter resolution, and this detail is improving at a breakneck pace.

Our ability to produce data is only rivaled by our ability to process that same data. Computer ecosystems are rapidly evolving and **Moore's Law**, which states that computing power will double roughly every 2 years, is alive and well! Advances in CPUs, GPUs, and data storage components, combined with improved coding languages and analytical methods, allow us to process data and make data-informed decisions faster than ever.

With all of this data and improved technology, organizations and individuals are looking for better and more efficient ways to derive meaning from data that may have once been treated as a byproduct of a technical process. This desire to find meaning in data has led to data science being one of the most in-demand skill sets of the 21st century.

In the introductory chapter of this book, *Geospatial Data Science with Python*, we'll begin by defining **Geographic Information Systems (GIS)**, data science, and geospatial data science. These definitions will lay the groundwork and begin to develop a common vernacular that will enable you, the data scientists and traditional GIS professionals, to work in harmony to solve some of the most complex and, dare we say it, fun problems of modern times.

In this chapter, we will cover the following topics:

- What is GIS?

- What is data science?

- What is geospatial data science?

What is GIS?

GIS stands for **Geographic Information Systems**. GIS are computerized systems used in the creation, collection, organization, analysis, and visualization of geospatial data. Geospatial data is a representation of the real world and it is rooted in **geography**. Geography is the study of the physical features of the Earth and its atmosphere, as well as how human activity impacts both. Human activity is looked at through many lenses, such as population distribution and land usage.

To represent the Earth in a GIS, you will leverage one of two data formats: **vectors** or **rasters**. *Figure 1.1* shows a stylized version of how real-world data can be represented in vector and raster formats. We'll define and discuss both of these terms in more detail in *Chapter 2, What Is Geospatial Data and Where Can I Find It?*

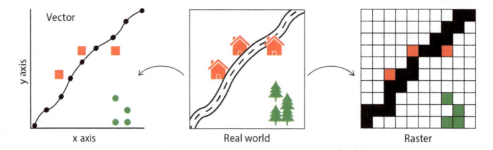

Figure 1.1 – Real-world data in vector and raster format

A typical GIS enables you to query and combine data assets in relation to the spatial relationship of each asset. This data is then visualized in the form of a static or interactive map or within a mapping application.

Geospatial data stored within a GIS comes in many different formats and from many different domains. A GIS used in local government may include information on the land parcels of local neighborhoods, the roads that run through that neighborhood, and the location of public service infrastructure, such as hospitals and fire stations. A GIS servicing a local weather station may include some of these assets, but will likely also include other types of data, such as real-time feeds of storm paths, rainfall totals, and wind speeds at various points around an area at various times. In *Chapter 2, What Is Geospatial Data and Where Can I Find It?*, we will focus more on various types of spatial data, their file structure, including shapefiles and GeoJSON, and some of the public sources in which spatial data can be found.

In your day-to-day life, you've likely used a GIS platform or an application more frequently than you may have realized. Take, for instance, **Google Maps**, which is arguably the most used GIS application in the world. Google Maps allows you to search for points of interest around you, such as a coffee shop or an auto mechanic, find directions to these points of interest, and also understand adverse conditions such as rush-hour traffic or roadworks that may impact your commute. There are many other forms of GIS applications out there, including applications that trace the route of an Amazon delivery vehicle as it approaches your home, applications that help you understand where public busses and transit hubs are located, and even applications that help monitor the spread of infectious diseases, as we mentioned in the preface to this book.

In addition to web and mobile GIS systems, there are also desktop-based, point-and-click GIS platforms that allow users to perform more complex spatial operations and analyses. These platforms are often used by specialized GIS practitioners who often have the title of geographer, GIS analyst, GIS engineer, or GIS specialist. These systems are used in a variety of different industries for different purposes. A GIS analyst in local government may use a desktop GIS platform to edit parcel boundaries within a town while a GIS analyst for a rail operator may use it to monitor the operation status and location of each railcar. The uses of GIS and the industries in which it is used are near limitless.

Typically, desktop GIS systems are provided by vendors, with the most dominant vendor in the space being **Esri**. As the dominant player in the GIS space, Esri's proprietary software integrates into numerous other applications with other vendors, including Microsoft and AutoCAD. In more recent versions of its software, Esri has also extended its application to work with many open source data science languages, such as Python and R, and **Integrated Development Environments** (**IDEs**), such as Jupyter Notebook. This book will focus on open source Python packages that do not require licensing. In *Chapter 4, Exploring Geospatial Data Science Packages*, we will cover packages including GeoPandas, PySAL, and GeoViews, along with many others you'll leverage in the case studies later in this book.

Now that you have an understanding of GIS, let's now define what data science is. As we define data science, hopefully, you'll begin to see how GIS and data science interact.

What is data science?

In the simplest terms, data science is the practice of deriving new insights from raw and disparate data assets and communicating those insights to stakeholders in a way that drives impact. The domain of data science combines facets of mathematics, statistics in particular, with computer science and industry- or domain-specific knowledge. Various sources and authors will define data science and the role of the data scientist differently, with some of these sources including soft skills such as communication and consulting as the fourth pillar of data science. These four pillars of data science are represented in *Figure 1.2*:

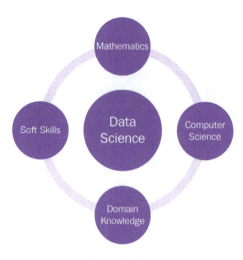

Figure 1.2 – Data science pillars

Each of these components alone can be tricky to master. It is important to recognize that most data scientists are not experts in all of these areas, but have foundational knowledge in these areas. This enables you to generate and communicate more robust and impactful insights with greater efficiency.

Before we dive deeper into the four components of data science, let's briefly discuss the data science pipeline at a high level. The pipeline can be thought of as follows:

1. Collecting
2. Cleaning
3. Exploring
4. Processing
5. Modeling
6. Validating
7. Storytelling

Often, the data science process is not performed in a linear fashion and instead can look something like the process displayed in *Figure 1.3*:

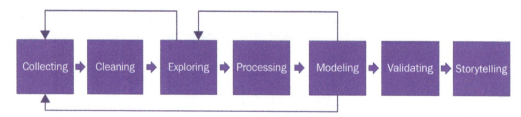

Figure 1.3 – Data science pipeline

While these are the general steps to be completed within a data science pipeline, every pipeline looks a bit different based on the problem you're trying to solve.

The skills needed at each of these steps also differ, which is why having knowledge across the four pillars of data science is so important. Now that we've talked about the data science pipeline, let's break down the four pillars into a bit more detail.

Mathematics

To be successful as a data scientist, knowledge of mathematics is required, as it is the underpinning of data science. Data science often focuses on taking raw data from the past, identifying patterns in said data, and making a prediction about what will happen in the future based on those patterns. In order to do this, applications of calculus, linear algebra, and statistics are necessary. However, you don't have to be an expert mathematician or statistician to understand how to identify and test the most suitable analytic method for the problem at hand.

Statistics is especially critical in the earlier stages of a data science process as you are sampling or developing a method for collecting data, as well as developing statistical hypotheses that will later be tested using the data you've collected. Statistics is also important as you begin to think through the types of algorithms that are appropriate for your analysis and then as you test each algorithm's related assumptions. *Chapter 6, Hypothesis Testing and Spatial Randomness*, will introduce you to hypothesis testing and the concept of spatial randomness, which is a critical hypothesis to test within the context of a geospatial data science workflow. In later stages of the data science process, calculus and linear algebra become more important, as they are the foundation of most algorithms.

Having knowledge of these subjects will allow you to understand the model you're developing, further refine its accuracy, and explain your model to end users in a comprehensible way. In *Part 3, Geospatial Modeling Case Studies*, of this book, we will focus on geospatial data science case studies that enact the full scope of the data science process and utilize a variety of algorithms in their solutions. Each case study will provide you with a greater perspective on how taking a geospatial data science approach to your analysis will provide you, and your stakeholders, with richer insights.

Computer science

Computer science is the next domain you'll need to understand to become successful in data science, as most jobs in this field will require some knowledge of data storage as well as programming skills in Python, R, SAS, **Structured Query Language (SQL)**, or another scripting language.

At the very beginning of the data science pipeline, you'll need to know where your data is stored or will be stored once it is collected. Relying on traditional file-based storage systems that store data as individual files is no longer suitable for day-to-day work in large enterprise settings. Often, a data scientist will need to use SQL to query data from a relational database such as Teradata, Oracle, or Postgres SQL, to name a few. SQL allows data scientists to query data from different tables and join the related data together based on common identifiers. Data scientists are often required to understand how to connect to a database, query the individual tables that make up the database, and export and transfer the data to the platform that will be used for further analysis, modeling, and visualization.

For geospatial data scientists, SQL also enables you to begin working with geospatial data through the use of spatial SQL. Spatial SQL enables you to perform many spatial operations with ease including point-in polygon intersections, spatial unions or joins, buffers, and Euclidian, or crow-flies, distance calculations. Users of more traditional, desktop-based GIS applications are often amazed at the spatial operations that can be performed, and repeated, with a few simple lines of code.

In the later stages of the pipeline, you'll need to know Python, R, or SAS in order to develop a machine learning or AI model on the data you've collected and explored in earlier phases. Toward the end of the data science pipeline, you'll then want to use these languages to visualize and interpret your results. For the purpose of this book, we will focus on the Python scripting language, as it is one of the more robust and extensible languages for data science in general, and in particular for geospatial data science. In *Chapter 4, Exploring Geospatial Data Science Packages*, we will focus on setting up your Python-based geospatial data science environment and provide you with an overview of the packages needed to perform various types of analysis and modeling.

For geospatial data science, you will often run into problems that require the use of large datasets or computationally intensive solutions. In each of these cases, knowledge of computer science skills becomes more important, as these problems can be solved better by leveraging the advancements in distributed (or parallel) computing and big data storage arrays. These environments allow you to break the process and data down into smaller chunks that can be distributed to multiple worker nodes in the parallel compute ecosystem. Breaking down the problem in this way can take a process that would have taken days, weeks, or even months on a single desktop and reduce the time to minutes or seconds.

Industry and domain knowledge

Having the technical skills to pull and analyze data and then develop a model is meaningless without knowledge relevant to the specific industry or domain in which the data science problem is rooted. In data science, there is an adage that states *garbage in, garbage out*, which often refers to bad data being used to generate a bad model.

Someone who doesn't have domain-related context will often pull data that isn't relevant or useful to solve the problem at hand. This bad data, when used to develop a model, will often not yield the insights that the stakeholder was expecting. To prevent pulling bad data, you'll often need to work hand in hand with stakeholders to understand the full context of the problem or issue you are trying to solve. Once this context is obtained, you'll be able to pull relevant data, understand the data in relation to the problem, and develop a perspective on the algorithm best suited for the individual situation. Industry- and domain-based knowledge are also necessary as you are developing and validating the results of your model in the later stages of the pipeline.

Soft skills

Often, a data scientist will not be working with other technical individuals or even conducting technical processes for 100% of their day. While a data scientist is required to have a strong technical background and understand the intricate nuances of programming and mathematics, it is rare that the end users or those being supported by a data scientist will have this same knowledge base. As mentioned in the previous section, a data scientist will need to rely on their stakeholders to understand and frame the problem at hand. This requires strong communication and collaboration skills to develop the working relationship. This relationship then becomes even more critical when a data scientist has completed their process and is working to interpret meaningful results. Often, a data scientist will also need to have strong consulting and influencing skills, especially in the business world, as they will need to influence stakeholders to implement and rely on the results of their technical processes.

To be a data scientist, you'll need to have a working knowledge across a wide array of topics as we've discussed in this section. However, data science is not a solo activity and you'll often be able to rely on others on your team or in the data science community to support and help you develop in the areas in which you're learning. Data science is a practice and we're all learning and growing every day.

Having developed an understanding of GIS and data science, you should now start to have an inkling about how they combine to form geospatial data science. We'll talk more about this powerful combination in the next section.

What is geospatial data science?

Geospatial data science lies at the intersection of data science and GIS as depicted in *Figure 1.4*:

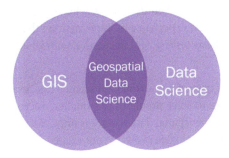

Figure 1.4 – Geospatial data science Venn diagram

Geospatial data science is a subclass of general data science that concentrates on geospatial data, its unique properties, and specialized techniques and computation methods necessary for deriving insights from this data. Instead of treating spatial data as another feature in a tabular dataset, spatial data science goes deeper into understanding why things are happening in a particular place and how they are related, or unrelated, to the things going on around it. Spatial data science focuses on identifying spatial relationships based on location, distance, and intersections between objects. We'll talk more about spatial relationships in *Part 2*, *Exploratory Spatial Data Analysis*, of this book, as we conduct exploratory spatial data analysis.

While this book will focus primarily on geospatial data science, that is, data science focused on data pertaining to the Earth, it is worth noting that the concepts can be expanded and translated to general spatial data science. Spatial data science focuses on where the datum is located and how it is related across space and therefore can be applied to problems at a microscopic scale, such as the location of atoms in your body, or problems much larger, such as the distance between asteroids in the main belt between Mars and Jupiter.

By looking at data while taking into account its spatial relationships, you will often find substantial improvements to existing or developing models. As mentioned at the start of this chapter, the domain of spatial data science is rapidly growing and new avenues for exploration are uncovered every day.

Summary

In this chapter, we defined the differences and commonalities between GIS, data science, and geospatial data science. As we discussed data science, we took a deep dive into the four pillars of data science, which include mathematics, computer science, domain and industry knowledge, and soft skills.

We also briefly discussed the stages involved in the data science process. Parts 2 and 3 of this book will provide you with more hands-on experience in implementing the data science process through exploratory data analysis, hypothesis testing, and in-depth data science use cases, covering a variety of topics and algorithms.

We also discussed how the principles of geospatial data science can be applied more broadly within the domain of spatial data science to solve problems at a smaller, microscopic level, as well as larger, astronomical scales. The power of geospatial data science is only starting to be realized as industries, data storage, and computing methodologies evolve. We're excited that you've decided to embark on this learning journey with us and are even more excited to see what you achieve in your journey to become a geospatial data scientist.

In the next chapter, we'll dive deeper into the world of geospatial data, which we briefly described in this chapter as being a representation of the real world in vector or raster format. We'll also spend time in the next chapter discussing the rich sources of open geospatial data.

2

What Is Geospatial
Data and Where Can I Find It?

To answer the first part of our question, **geospatial data**, in simplest terms, is data that has a geographic component—that is, a component that ties the data to a point on, or adjacent to, the Earth's surface. To answer the second part of our question, geospatial data is quite literally all around you.

There are large volumes of geospatial open data that is collected, maintained, and released by public entities such as government agencies or **non-governmental organizations (NGOs)** as well as private corporations. The availability of geospatial data within the open data ecosystem has led to the rise of robust data standards that were developed and are actively used by the geospatial community. The development of community standards is talked about briefly in this chapter as it relates to data formats and publications.

In this chapter, you'll learn about the following topics:

- Static and dynamic geospatial data

- Geospatial file formats

- Introducing geospatial databases and storage

- Exploring open geospatial data assets

Static and dynamic geospatial data

Geospatial data can typically be viewed as **static** or **dynamic**. Static geospatial data is data that does not change over a short-term time period. This data can include things such as the epicenter of an earthquake, the location of a store, or the number of college-educated adults. Dynamic geospatial data, in contrast, can change in real time. This data can include the location of a shopper within a shopping mall, the position of a bike courier delivering food, or the spread of an infectious disease such as the SARS-CoV-2 virus that caused the COVID-19 pandemic. Dynamic geospatial data is often referred to as **spatiotemporal data**, or data relating to both space and time. Static and dynamic geospatial data comes in two formats: **vector** and **raster**. We'll discuss them in detail in the following section.

Geospatial file formats

As mentioned at the start of this chapter, geospatial data is data with a geographic component. This geographic component is often a latitude and longitude coordinate that is collected via a **global positioning system** (**GPS**). A geographic component can also be derived from an address using a process called **geocoding**. However, there are also many other geographic components that can relate tabular, or attribute, data to standard administrative geographies. We'll talk more about administrative boundaries at length later in our discussion on vector data. It is also worth noting that geospatial data is a subset of **spatial data** or data that is related to a point in some broader study space.

Vector data

Vector data is not unique to the field of **geographic information systems** (**GIS**) or geospatial data science and has applications in many digital mediums. When talking about vector data in GIS or geospatial data science, we are talking about data that represents real-world features. The foundation of vector graphics is a vertex or point that is typically denoted by an X and Y coordinate. You can think of this point as the location of your favorite ice cream shop or the location of a fire hydrant. In a GIS, X represents longitude and Y represents latitude, and these coordinates are relative to a spatial reference, or projection. We'll talk more about spatial projections in *Chapter 3, Working with Geographic and Projected Coordinate Systems*.

If you have two or more vertices, they can be connected by paths to form polylines. In a GIS, you may have tens to hundreds of vertices that, when combined and connected, represent a highway connecting major cities. Finally, a series of polylines can be connected to form a polygon. These polygons could represent a building footprint or an administrative boundary such as a state or country. Polygons can also have interior vertices and polylines that carve out internal sections and form multipart polygons.

A vector data example

To better understand vector data and how it represents the real world, let's review a few real-world visualizations. *Figure 2.1* shows a high-level map of Manhattan in New York City:

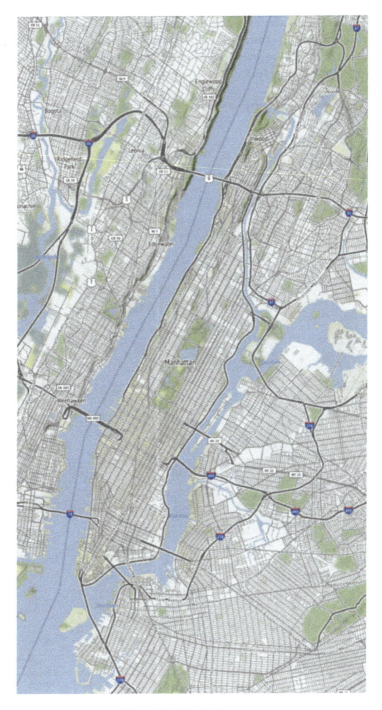

Figure 2.1 – Manhattan map

As you can imagine, Manhattan is a bustling island with numerous amenities including fun attractions for people to see, roads to get them there, and buildings for them to stay in while they're in town. *Figure 2.2* shows a map of Manhattan overlayed with points that represent area attractions such as the Statue of Liberty and the Central Park Zoo:

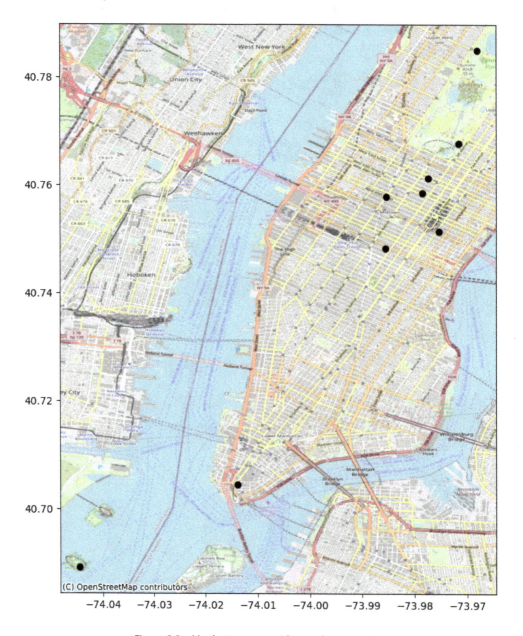

Figure 2.2 – Manhattan map with popular attractions

Now, we'll zoom in on the Upper East Side, Central Park, and Upper West Side neighborhoods. Here, we can see the many roads that run throughout these neighborhoods, represented as lines on the map:

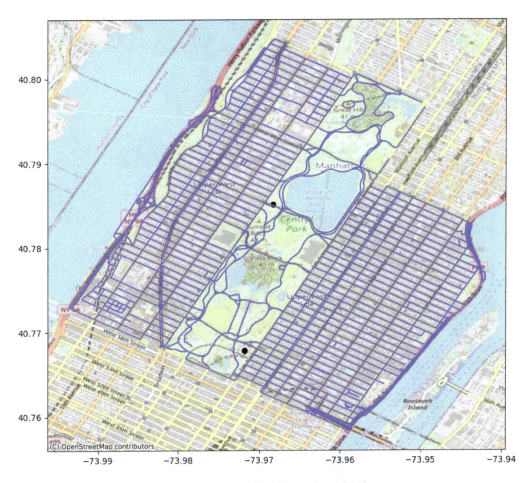

Figure 2.3 – Map of roads near Central Park

Next, we'll overlay the buildings within these neighborhoods, which are displayed in *Figure 2.4*. These buildings are represented as polygons on the map:

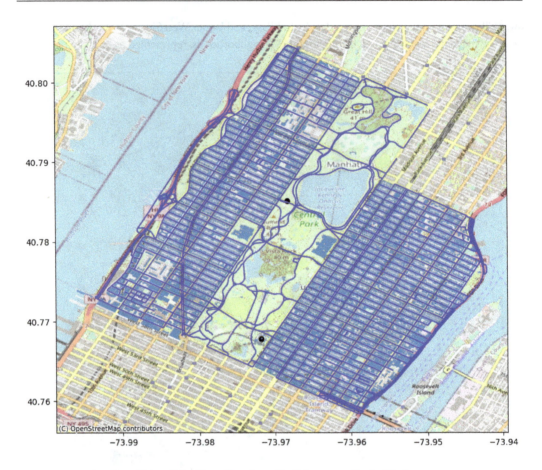

Figure 2.4 – Map of roads and buildings near Central Park

Lastly, *Figure 2.5* shows a map of the roads and buildings near Central Park overlayed with the administrative boundaries for the Upper East Side, Upper West Side, and Central Park neighborhoods. This administrative boundary is represented as a polygon on the map:

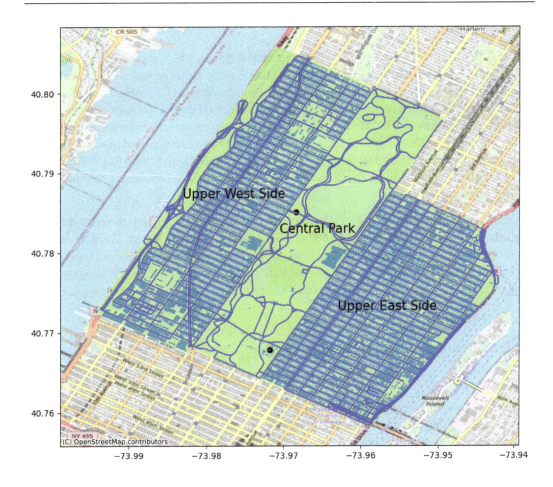

Figure 2.5 – Map of roads and buildings near Central Park with neighborhoods

As we discussed previously, vector data is a representation of the real world in the form of points, lines, and polygons. Points are the first building block and can be used to represent things such as attractions. When points are connected, they form lines that can be used to represent roads. Finally, when lines are connected, they form polygons that can be used to represent building footprints or administrative boundaries such as neighborhoods.

Other vector data uses

X, Y, and Z coordinate data can also be used to create **point clouds**, which are becoming more and more ubiquitous as **Light Detection and Ranging**, or **LiDAR**, technologies become more mainstream. This is most notable in the domain of self-driving cars, which use LiDAR for object detection and avoidance. While LiDAR is one mechanism for creating point cloud data, it is not the only means of creation. Point clouds can also be created from photographs via a process called **photogrammetry** to convert overlapping 2D images into 3D models of objects. Photogrammetry is used in surveying and mapping both here on Earth and up in space. The recently launched James Webb Space Telescope, which will replace the Hubble Space Telescope, will utilize specialized instruments and photogrammetry to uncover new reaches of space.

Now that you understand vector data a bit better, we can begin introducing you to vector data file formats that you may encounter during your work as a geospatial data scientist. There are far too many file formats to go into in detail in this book, so we'll focus on those that you are most likely to encounter. Details on file formats not covered in this book can be found on *GIS Geography* by visiting `https://gisgeography.com/gis-formats/`.

Vector file formats

In the next few sections, we'll discuss the most popular vector file formats.

Shapefile

A shapefile is a geospatial file format that was originally developed by Esri as an open specification data storage format that supports interoperability between Esri's ArcGIS platform and other GIS systems. Given the ubiquity of geospatial data, the shapefile has become the mainstream file format for storing and sharing vector data. In addition to storing the spatial geometry of points, lines, and polygons, the shapefile format also stores attribute information related to those features.

A shapefile is a multipart file format that requires four main parts:

- `.shp`—The geometry of a point, line, or polygon feature
- `.shx`—The index of a feature
- `.dbf`—Attribute data that stores columnar variables related to their features
- `.prj`—Projection metadata that utilizes **well-known text** to store information related to the **projection and coordinate reference system**

A shapefile can also include several other parts, including `.sbn`, `.sbx`, `.fbn`, `.fbx`, `.ain`, `.aih`, `.cpg`, and `.qix`, to name a few. For more information about the multitude of possible optional subcomponents of a shapefile, visit `https://en.wikipedia.org/wiki/Shapefile`.

> **Important note**
>
> You should be aware of the following points when it comes to creating and working with shapefiles:
>
> - A shapefile will become corrupted if any of the four required parts are deleted from the shapefile.
>
> - Neither the `.shp` nor the `.dbf` subcomponents of a shapefile can exceed 2 GB in size. This limit often makes shapefiles an inconvenient storage format for larger data assets.

The United States Census Bureau maintains a specialized shapefile format called **Topologically Integrated Geographic Encoding and Reference system**, or **TIGER**, files. TIGER files do not contain attribute data that is collected by census products. Unlike the decennial census or the **American Community Survey (ACS)**, they map features including standard census geographies such as census tracts. They are also used to map other public geographic features such as roads and railroads. TIGER files contain **geographic entity codes**, or **GEOIDs**, which can be used to link them together with other census data products also containing GEOIDs.

GeoJSON

Geographic JavaScript Object Notation (GeoJSON) is the geographic sibling of the more common **JavaScript Object Notation (JSON)**. GeoJSON formats are mostly used for web-based mapping as web browsers understand how to interpret JavaScript. GeoJSON file formats store the coordinates of the geometry as well as the columnar attribute information related to those geometries as text within curly braces: { }. This file format can easily be read by any text-based file editor as well as web-based tools for working with JSON data—for example, CodeBeautify's JSON Viewer.

KML

Keyhole Markup Language (KML) is a file format used to store and display geographic data that was created by Google. Google transitioned the KML file format to the **Open Geospatial Consortium (OGC)** to maintain and evolve into a standard format for displaying GIS data on web-based and mobile-based 2D maps as well as 3D Earth browsers.

KML is an XML language that is primarily focused on geographic data visualization, including annotating maps and images. KML is not just concerned about displaying data but is also focused on assisting the end user in their navigation by providing them with context on what to look for and how to get there.

For more information on the KML file format, visit the OGC at `https://www.ogc.org/standards/kml/` or *Google Developers* at `https://developers.google.com/kml/documentation/kmlreference`.

OSM

The OSM file format is an XML-based file format that was created to store and easily distribute geospatial data by OpenStreetMap. OpenStreetMap is one of the largest crowdsourcing communities for geospatial data. The OSM file format is a collection of vector-based features from this crowdsourcing community. We'll talk more about OpenStreetMap and its data catalog later on in this chapter.

Raster data

In comparison to vector data, which uses points, polylines, and polygons to model real-world objects, raster data approaches modeling the real world differently. Raster data is any picture data that is composed of uniform cells or pixels. Each cell within a raster data file is typically square, but raster data cells can take other shapes as well. *Figure 2.6* illustrates a raster data file:

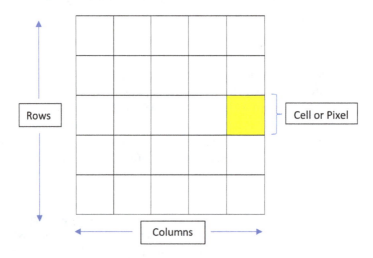

Figure 2.6 – Raster data file

Raster data takes the form of a matrix of cells, or pixels, organized in a uniform row-by-column architecture. In geospatial data, each cell is geolocated to a specific point on the Earth's surface, and the value of each cell represents a measurement at that location. Raster data is typically used for continuous data, which cannot easily be formatted as vector data.

Take, for example, data on land usage in a local agricultural district. Vector data can easily be used to distinguish the parcel boundaries denoting each farm as well as the location of windmills and wells, but it is not as useful in distinguishing what type of crop is growing in certain areas of the land compared to what obstruction is occupying other areas. Consider the following land use map:

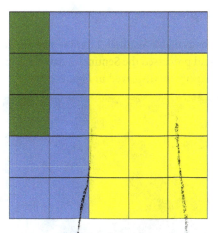

Figure 2.7 – Raster land use map

In *Figure 2.7*, the yellow pixels could represent land that is being utilized to grow wheat. The blue areas that wrap around the wheat may represent a natural boundary in the form of a river and pond. Finally, the upper-left pixels shaded in green may represent an uncleared forest that is not yet suitable for crop production.

Oftentimes, users of a GIS will use raster data as a background layer underneath vector data to provide more context to the vector data. In your day-to-day life, you have likely experienced this when you've opened up Google Maps, which can overlay vector street network and **points of interest** (**POIs**) data with raster satellite data. In our land coverage example, you could combine the raster data with vector data denoting the points of windmills and wells, as we discussed previously. You can see its representation in *Figure 2.8*:

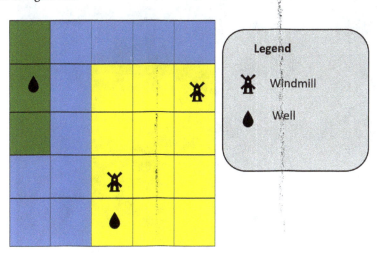

Figure 2.8 – Raster land use map with vector data

Real-world raster file example

The European Space Agency makes available Sentinel-2 remote sensing data via the Copernicus Open Access Hub. We've pulled down and processed the Sentinel-2 data for Sandusky, Ohio. The red, green, and blue bands of the satellite imagery are displayed in *Figure 2.9*:

Figure 2.9 – Sandusky, Ohio – Sentinel-2 RGB bands

It is hard to tell from this view what the imagery is displaying. *Figure 2.10* displays the true-color adjusted imagery:

Figure 2.10 – Sandusky, Ohio – Sentinel-2 true color

From the preceding screenshot, you can begin to make out parts of Sandusky Bay, the Resthaven Wildlife Area, and the Cedar Point amusement park. We'll talk more about satellite imagery in *Chapter 4*, *Exploring Geospatial Data Science Packages*.

Raster file formats

As with vector data, there are several raster file formats that you may run into at some point in your journey as a geospatial data scientist.

GeoTIFF

The GeoTIFF file format was created in the late 1990s and is based on the **Tagged Image File Format** (**TIFF**). TIFF file formats are widely used in image-manipulation applications and numerous other types of applications. GeoTIFF is an evolution of the TIFF file format in that it allows for the addition of georeferencing information within the image, thus allowing for geographic metadata to be accessible along with the image file. These image files are typically sourced through satellite imagery, aerial photography, and digital maps. TIFF image files support both RGB and CMYK color spaces.

The metadata included in the GeoTIFF file format includes information relating to the following:

- The vertical and horizontal components of the raster
- The **coordinate reference system** (**CRS**) that the data is based on
- The spatial extent and spatial resolution
- Rules for how to project the raster data into a 2D digital medium

The OGC has set forth the *OGC GeoTIFF 1.1* format standard to help formalize pre-existing standards used by the GeoTIFF community and help to further develop the format based on new needs of the community and changes in technologies.

JPEG

Joint Photographic Experts Group (**JPEG**) is an open source standard image format for containing lossy and compressed image data. The JPEG file format is not unique to geospatial data as it is one of the most common image file formats. JPEG files also did not allow for the inclusion of georeferenced metadata until the release of the JPEG 2000 format.

PNG

Portable Network Graphics (**PNG**) is another popular image format that supports georeferenced metadata. In comparison to the JPEG file format, PNG files support both lossy and lossless compression and make use of 24-bit images.

In this section, we've discussed the various file formats used to store vector and raster data. In the next section, we'll introduce you to geospatial databases and storage software.

Introducing geospatial databases and storage

This section will introduce you to geospatial databases and storage software. Let's begin the discussion with PostgreSQL and PostGIS.

PostgreSQL and PostGIS

PostgreSQL is an open source, object-relational database system that uses **Structured Query Language (SQL)**. PostgreSQL runs on all major operating systems and has been praised for its architecture, scalability, reliability, and extensibility. While PostgreSQL isn't spatially enabled by default, it can be spatially enabled through the use of the PostGIS database extender.

PostGIS is a project of the **Open Source Geospatial Foundation** (**OSGeo**). PostGIS adds spatial operations such as distance, area, union, and intersection, as well as spatial geometry, to the standard PostgreSQL database. If you've worked with a standard database in the past, then you're likely familiar with its row-by-column storage structure as well as its ability to join different sets of data, or tables, together based on a common ID or key. In a standard database, you can efficiently query numerical, text, timestamp, or image-based data that can answer questions such as: *How many purchases occurred at store X on date Y? How many people logged in to my mobile app during the month of April?* PostGIS extends the questions you can answer through spatial data types and operations such that you can efficiently answer questions such as: *How far away is the nearest pharmacy from X address? What is the size of this census tract?*

PostGIS has three main capabilities:

- Multi-dimensional spatial indexing, allowing for the efficient processing of spatial functions
- Spatial functions written in SQL to query spatial operations and perform spatial operations. These functions include the following:

  - Generation functions that generate new geometries
  - Retrieval functions that can retrieve spatial data and measurements of spatial geometry
  - Management functions that allow you to manage data stored in the database
  - Conversion functions for converting external data to spatially enabled data and vice versa
  - Comparison functions that compare two features and their geometries

- Spatial geometries including points, lines, and polygons

To properly do justice to PostgreSQL and PostGIS, this would require an entire book, and for brevity, we've decided not to go further on this topic. With that being said, there are numerous detailed overviews of both topics. For PostgreSQL, check out `https://www.postgresql.org/about/`, and for PostGIS, visit `https://postgis.net/workshops/postgis-intro/`.

ArcGIS geodatabase

The ArcGIS geodatabase is a proprietary database created by Esri that is used within its ArcGIS suite of products. The geodatabase is a collection of spatial datasets represented as Esri feature classes, tables, or shapefiles that are stored in a common folder structure on your system or in a relational database such as Oracle, PostgreSQL, or an IBM Db2 instance. Geodatabases can vary in size serving an individual or small user base as a file geodatabase or an entire enterprise when deployed on a database server.

Geodatabases can be assessed from within the ArcGIS ecosystem or by leveraging specialized Python packages, including osgeo, fiona, and geopandas. We'll talk in more detail about these Python packages and others in *Chapter 4, Exploring Geospatial Data Science Packages*.

Exploring open geospatial data assets

As we mentioned at the start of this chapter, over the last two decades, geospatial data has become highly available to the public due to the rise of internet-connected technologies as well as the rise of numerous open source communities. In this section of the book, we're going to cover open geospatial data assets that are provided by government agencies, NGOs, citizen scientists, and some amazing geospatial communities.

Human geography

Human geographic data or anthropogeographic data is a branch of geographic data that deals with humans and their relationship to the area around them.

United States Census Bureau data

Within the United States, the Census Bureau is one of the best sources of **geo-demographic data**. Geo-demographic data is location-based data that segments individuals based on their locale as well as their demography. The Census Bureau collects this data via various programs including the decennial census as well as the yearly ACS, the biennial **American Housing Survey** (**AHS**), and the yearly **American Business Survey** (**ABS**). The following subsections provide more details on census data structure and its **unique IDs** (**UIDs**).

GEOIDs

As mentioned previously, census data products include geographic entity codes or GEOIDs. There are two primary types of GEOIDs: **Federal Information Processing Standards** (**FIPS**) codes and **Geographic Names Information System** (**GNIS**) codes.

FIPS codes

The Census Bureau started publishing FIPS codes with its data products over 30 years ago. FIPS codes are assigned in alphabetical order by geography name for states, counties, **core-based statistical areas**

(**CBSAs**), and county subdivisions. The FIPS code system typically ensures that smaller geographic units are unique within larger geographic units. That is to say that FIPS codes for states are unique within a country and counties are unique within a state. Due to this structure, most census geographies nest neatly inside one another, as shown in *Figure 2.11*.

For some data products that use FIPS codes, they do not sit neatly next to the standard hierarchy of census geographic entities. Some of these areas include urban areas, CBSAs, state legislative districts, and ZIP code tabulation areas, to name a few. In the following diagram, these special cases are shown as lines to the side of the standard hierarchy:

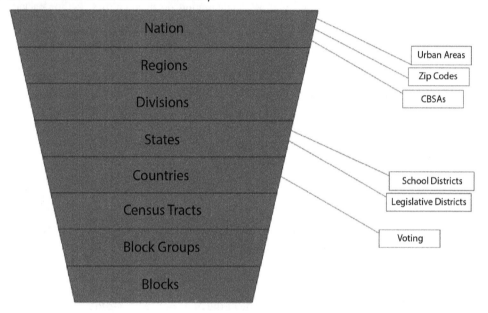

Figure 2.11 – Census geographic data hierarchy

FIPS codes are 15 digits long, with the first 2 digits representing the state, the next 3 digits representing the county, the next 6 digits representing the census tract, the next digit representing the block group, and the final 3 digits representing the individual census block.

FIPS code 390017701001000 can be read as follows:

- 39—Ohio (black)
- 001—Adams County (blue)
- 770100—Census tract within Adams County (red)
- 1—Block group within tract 770100 (green)
- 000—Block within block group 1 (purple)

The census hierarchy just described is displayed in *Figure 2.12*:

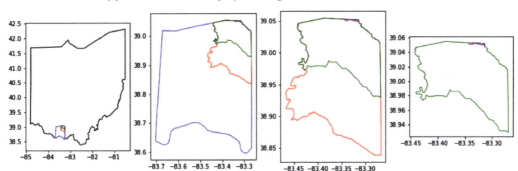

Figure 2.12 – Ohio census geographic data hierarchy

GNIS codes

Unlike FIPS codes, geographic features that utilize GNIS codes do not have a nesting relationship with one another. GNIS codes are assigned sequentially based on when they were entered into the GNIS database. Many geographic features that represent physical or culturally important features are represented in the GNIS database, including airports, beaches, hospitals, post offices, schools, populated places, and cemeteries, to name a few.

Data products

Census geodemographic data products can easily be downloaded from `https://data.census.gov/cedsci/`. This geodemographic information can be highly informative in numerous problem spaces. Later on, in *Section 3* of this book, we'll leverage census data within our hands-on case studies.

OpenStreetMap

As mentioned previously, the OpenStreetMap Foundation maintains data sourced from OpenStreetMap's community of contributors. **OpenStreetMap (OSM)** data includes all of the relevant information that you'd find on a physical map, including road networks, POIs, buildings, and landmarks.

The OpenStreetMap road network dataset is a routable dataset that can be leveraged in Python. By leveraging OSM data, we can create new geospatial features based on distance calculations generated by driving, walking, or public transit modes. OSM street network data is also useful when it comes to solving **vehicle routing problems (VRPs)** or shortest path problems. Later on, in *Part 3, Geospatial Modeling Use Cases*, you'll be given some hands-on experience using OSM data and numerous Python packages to solve these and other problems.

Finally, OSM data can also be used as a **reference layer** within mapping products. Reference layers can be anything from road networks, landmarks, and physical geographies such as lakes and rivers, to administrative boundaries. These reference layers provide end users of the mapping product with an easy way to orient themselves to the geography represented by the map.

United Nations Environment Programme geodata

The **United Nations Environment Programme** (**UNEP**) maintains a rich catalog of human geographic data that covers geographies across the globe. The data included in its catalog represents features such as light pollution, human pressures on protected or endangered species, and access to healthcare facilities. At the time of writing this book, there were 205 unique data assets within the catalog. To access this data, visit `https://datacore-gn.unepgrid.ch/geonetwork`. Data can be accessed via the search function, categories, or individual data tiles.

University of Wisconsin Center for Sustainability and the Global Environment

Scientists at the University of Wisconsin Center for **Sustainability and the Global Environment** (**SAGE**) produce and maintain data assets, maps, models, and software built upon human geography. Most of the data assets in their catalog are grid-based datasets that represent topics such as air quality, urban expansion, and crop calendars. To view SAGE's data catalog, visit `https://sage.nelson.wisc.edu/data-and-models/datasets/`.

CIA World Factbook

The United States **Central Intelligence Agency** (**CIA**) maintains a data catalog called the World Factbook that provides intelligence regarding governments, people, geography, transportation systems, militaries, and terrorism for 266 entities across the world. Its data catalog is broken out by countries, world regions, and world oceans, allowing end users to easily navigate and find data relevant to their problem space. Visit `https://www.cia.gov/the-world-factbook/` for more information on the CIA World Factbook.

The next section will introduce you to physical geography-focused geospatial data.

Physical geography

Physical geography is a branch of geospatial data that represents the physical, or natural, environment. This includes things such as weather, climate, land formations, plants, and natural phenomena such as earthquakes and tsunamis.

United States Geological Survey

The **United States Geological Survey** (**USGS**) maintains a rich data catalog of both real-time and historical physical geography. Real-time data provided by the USGS includes active monitoring of earthquakes, landslides, volcanoes, wildfires, and geomagnetism.

The USGS, in partnership with the **National Aeronautics and Space Administration** (**NASA**), publishes and maintains **Landsat** data. Landsat data has been published since 1972 and is sourced from continuous images captured by NASA's Earth observation satellites. In February 2022, the USGS started to release data from the Landsat 9 satellite. The Landsat 9 satellite includes two instruments: the **Operational Land Imager 2** (**OLI-2**) and the **Thermal Infrared Sensor 2** (**TIRS-2**). These new instruments produce raster data that can differentiate 16,384 shades of given wavelengths compared to 4,096 shades in the Landsat 8 satellite.

The USGS has two collections of **Analysis Ready Data** (**ARD**):

- **Collection 1 ARD**: Based on Landsat 4-5, 7, and 8 data from 1982 through 2021
- **Collection 2 ARD**: Based on Landsat 4-6, 7, and 8-9 data representing products from 1982 through to the present day

To learn more about the data provided by the USGS, visit `https://www.usgs.gov/`.

National Aeronautics and Space Agency (NASA)

In addition to its partnership with the USGS, NASA also provides Earth observation data via its Earthdata portal. Data assets in the Earthdata portal are sourced from satellite and airborne platforms, model outputs, and field activities. The data covers a range of topics, including atmosphere, calibrated radiance and solar radiance, cryosphere, human dimensions, land, and oceans.

The Earthdata Search function at `https://search.earthdata.nasa.gov/search` provides you with an easy way to discover and filter data based on platform, instrument, data format, and resolution. To download data and use data in your workflows, you will need to register for a free account by visiting `https://urs.earthdata.nasa.gov/users/new`.

OpenTopography

OpenTopography is an initiative based out of the **San Diego Supercomputer Center** (**SDSC**) at the University of California, San Diego. OpenTopography is on a mission to democratize access to high-resolution topographical information that is acquired via LiDAR and other technologies. It is also working hard to foster interaction and knowledge exchange within the Earth science community.

There are two ways of discovering data provided by OpenTopography. The Find Topography Data portal is an interactive mapping application that lets you discover topographical data by navigating the map and clicking on the geography you're interested in. In addition to this map-based approach, you can also query the data catalog directly. Data within OpenTopography's catalog covers the United States and select international locales. For more information on OpenTopography, visit `https://opentopography.org/start`.

The next section will introduce you to country- and area-specific geospatial data.

Country- and area-specific data

Most of the previous data sources were centered on the geography of the United States or data providers within the United States. As you can imagine, there is a plethora of open geospatial data for countries and areas across the globe. In this section, we'll cover a few of those data assets.

Open Government – Canada

Canada's Open Government portal is similar to the United States Data.gov portal and contains a variety of open data, including geographic data. Data included in this portal pertains to Canada's population, satellite-based land coverage data, hydrology, and data sourced from radar systems. Canada previously had a data source called GeoGratis that has since merged with the Open Government portal. Topographic data previously available in GeoGratis can now be found in Open Government. Visit `https://open.canada.ca/en` to learn more.

India Earth observation

The Indian government hosts a platform to view and download Earth observation data from satellite arrays. Data within this portal comes from various satellites, including SCATSAT-1 and OCEANSAT-2. Data products focus on land coverage, oceans, and India's physical terrain. Visit `https://bhuvan-app3.nrsc.gov.in/data/download/index.php` to learn more.

The South African Risk and Vulnerability Atlas

The **South African Risk and Vulnerability Atlas (SARVA)**, found at `https://sarva.saeon.ac.za/about/`, is an open science portal that provides access to geospatial data as well as infographics. The focus of these data assets is on natural and human activity-based hazards facing South Africa. Data assets include information on urbanization, historical rainfall, temperature changes, biodiversity, and healthcare.

Summary

In this chapter, we've covered a variety of topics related to geospatial data. First, we introduced you to the two types of geospatial data: vector and raster data. Vectors represent physical geography via points, lines, and polygons. Rasters represent physical geography as a continuous grid of pixels.

We then introduced you to the various types of file formats you may encounter when working with vector- or raster-based data. When it comes to vector data, we covered shapefiles, GEOJSON, and KML files. For raster-based data, we introduced you to GeoTIFFs and georeferenced JPEGs, and PNG files.

After our discussion of geospatial data formats, we walked through two different types of geospatial databases: the PostGIS-enabled PostgreSQL database, as well as the geodatabase provided by Esri.

We concluded the chapter with broad coverage of open source geospatial data on topics of human geography, physical geography, and country- and area-specific data. Given the variety of geospatial data available, this serves only as an introduction, and we encourage you to discover the geospatial data assets available in your local area or a topic that interests you.

In the next chapter, we'll introduce you to the many Python packages that you'll leverage in your work as a geospatial data scientist.

3

Working with Geographic and Projected Coordinate Systems

Contrary to what some may believe, the Earth is not flat. In fact, early Greek people began theorizing that the Earth was not flat as early as 500 BC when Pythagoras proposed that it was round. Pythagoras derived his theory that the Earth was round based on the sunlight that is projected off the Moon's surface. He noted that the line between the light and dark zones of the Moon is curved. Thus, he concluded that the Moon was a sphere, and thus, all celestial bodies must also be spheres.

Sometime between 384 and 322 BC, Aristotle, another Greek philosopher, added additional evidence supporting the spherical Earth model. His evidence came from studying the consistency in the curvature of lunar eclipses and noted that only a spherical Earth could produce such consistency.

Aristotle also observed that, as the Greek people sailed for new lands in the north and south, there were changes in the altitude of constellations and changes in the visibility of ships as they crossed the horizon. He posited that if the Earth was flat, then individuals at two ends would share the same horizontal plane, but in reality, as a ship crossed the horizon, it would disappear from the perspective of an individual at another point. With this logic, Aristotle rejected the hypothesis that the Earth could be flat and concluded that it must, indeed, be spherical. This observation is depicted in *Figure 3.1*.

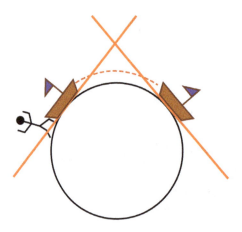

Figure 3.1 – Depiction of Aristotle's ship observation

> **Note**
>
> Galileo Galilei is credited with the creation of the **scientific method** and experimental **hypothesis testing** about 1,800 years after Aristotle's hypothesis of a round Earth. Aristotle's hypotheses are often referred to as informed hypotheses, compared to Galilei's hypothesis testing utilizing null and alternative hypotheses. We'll talk more about hypothesis testing in *Chapter 6, Hypothesis Testing and Spatial Randomness*.

Given that the Earth is round, the best models of a round Earth come in the form of globes. However, traveling with a globe is quite difficult, and we often want to represent the round, three-dimensional Earth on a two-dimensional surface, such as a paper map or on the screen of a laptop or mobile device. In order to do this, we need to understand **map projections** and the **coordinate reference system** (**CRS**). We introduced you to this topic in the previous chapter during our discussion on geospatial data and the geographic components of geospatial data, the **latitude** and **longitude** coordinates.

By the end of this chapter, you'll have learned about the following key topics:

- Geographic coordinate reference systems and spatial projections

- How to project and reproject data utilizing Python

- How to avoid some common pitfalls when dealing with spatial projections

Technical requirements

In this chapter, you'll leverage the `Chapter 3 - GCS and PCS_vf.ipynb` Jupyter notebook, which you'll find in the book's GitHub repository at `https://github.com/PacktPublishing/Applied-Geospatial-Data-Science-with-Python/tree/main/Chapter03`.

Exploring geographic coordinate systems

A **geographic coordinate system** (**GCS**) leverages a three-dimensional spheroid surface to define locations on or above the Earth's surface. These locations are identified based on their longitude and latitude values. In most geographic information systems, you will see longitude represented by an X value and latitude represented by a Y value. Points that are above the Earth's surface have a Z score representing their altitude above the surface. Longitude and latitude are angles that span outward from the center of the Earth. The angles are often measured in degrees (°) of the circle, where the entire circle is 360 degrees.

The network of latitude and longitude lines is known as a **graticule**, as shown in the following diagram:

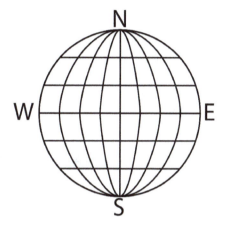

Figure 3.2 – A graticule network

Figure 3.3 shows the Earth with a graticule network based on a GCS.

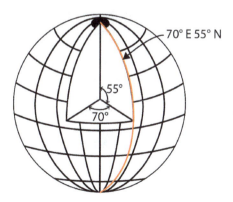

Figure 3.3 – The Earth with a GCS

The lines that run north to south, or vertically, in the GCS have a constant longitude value for the entire length of the line. These lines are called **meridians**. The meridians form circles around the Earth and intersect at the North and South Poles. The central meridian is called the **prime meridian**. Greenwich, England, is the most commonly known prime meridian and is also known as the **Greenwich meridian**. Other cities, including Paris, have also been used as prime meridians. The Paris prime meridian is called the **Paris meridian**. *Figure 3.4* depicts the prime meridian and the Earth's poles.

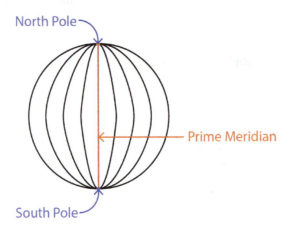

Figure 3.4 – The Earth with longitude lines

Meridians that are east of the prime meridian have values ranging from 0° to +180°, up to the **antipodal meridian**. Meridians that are west of the prime meridian have values ranging from 0° to -180°, up until the antipodal meridian. The antipodal meridian has a value of both 180° W and 180° E, while the prime meridian has a value of 0° E and 0° W.

> **Note**
>
> The Greenwich meridian was established by the Royal Observatory, which is located in Greenwich, England. In the 1880s, the International Meridian Conference took place in Washington, D.C., with delegates from numerous countries voting to approve the Greenwich meridian as the single meridian of the international community. French delegates abstained from voting as French map makers preferred the Paris meridian, and they continued to use the Paris meridian in their maps after the International Meridian Conference. In 1984, what is now the International Earth Rotation and Reference Systems Service established the **IERS Reference Meridian**, which is now leveraged by global positioning systems in the WGS84 projection. The IERS Reference Meridian better tracks the 0° longitude based on the center of the Earth's mass relative to tectonic shifts.

The lines that run east to west, or horizontally, in the GCS have a constant latitude value and are also referred to as parallels. These lines are equidistant from one another, and they form circles that are concentric around the Earth. The parallel that divides the Earth in half is called the equator. The equator has a latitude value of 0° and is halfway between the North and South Poles. Latitude lines that are north of the equator have values ranging from 0° to +90°, while latitude lines south of the equator have values ranging from 0 to -90°. *Figure 3.5* depicts the Earth with latitude lines.

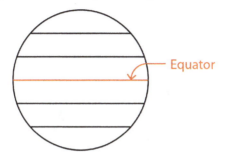

Figure 3.5 – The Earth with latitude lines

Let's now move on to cover GCS versions next.

Understanding GCS versions

There are many different models of the Earth's surface, and therefore, there are many different GCS versions. The reason that there are many GCS versions is that each GCS is a model of the Earth's surface and not a perfect representation of the Earth. This is because the Earth's surface is highly variable, from high mountains, such as Mount Everest, to very deep trenches, such as the Marianas Trench. The Earth is also constantly spinning, which makes the poles of the Earth pull slightly closer to the center of the Earth while the equator stretches a bit further out from the center. Instead of being a perfect sphere, the Earth's true shape is closer to an **oblate spheroid** or a squashed spherical object with a smaller polar circumference than an equatorial circumference. Given the imperfections in our models of the Earth's surface, different GCS versions are going to be more or less accurate, depending upon the geography that your analysis is focused on.

Let's now talk about a few of the most common geographic coordinate systems.

World Geodetic System 1984

World Geodetic System 1984, also known as **WGS 1984**, is the GCS that is standard for the **Global Position System (GPS)**. WGS 1984 utilizes Earth's center mass as the coordinate system origination. **Geodesists**, those who measure and monitor coordinates on the Earth's surface, believe that WGS 1984 has an error of less than two centimeters.

WGS 1984 is a global ellipsoid model, which is an extremely precise model of the Earth's surface. **Global ellipsoid models** were only able to be developed after GPS went mainstream and computer technology and satellite systems improved. These models are created by GPS satellites, which bounce radio waves off the Earth's surface from multiple angles. These radio waves are then processed via **trilateration** to accurately pinpoint a location on the Earth's surface. *Figure 3.6* depicts an array of three satellites used to triangulate a point on the Earth's surface.

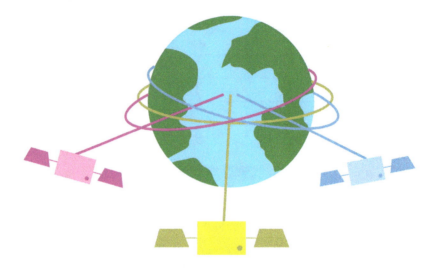

Figure 3.6 – GPS trilateration

WGS 1984 is currently the unified global ellipsoid model and is the standard for GPS coordinates. Before WGS 1984 was invented, there were other global ellipsoid models that had varying degrees of inaccuracy, including WGS 1972 and GRS 80. WGS 1984 is leveraged as the de facto GCS in most web and mobile mapping applications, including Google Maps, Waze, and Bing Maps.

> **Note**
>
> Each GCS has an identifier called a **well-known ID,** or **WKID** for short. The WKID for WGS 1984 is 4326. When working with GCS and **projected coordinate systems (PCSs)** in Python, it is the WKID for the GCS or PCS that you will reference as a parameter in your code.

GCJ-02

GCJ-02 is a GCS that was approved by the Chinese government's State Bureau of Surveying and Mapping in 2002 to protect national security interests. GCJ-02 is based on WGS 1984, but it layers on an obfuscation algorithm that adds noise to the actual location on the Earth's surface. The amount of noise added is random and adds an offset to both the true latitude and longitude. The maximum amount of offset that is added is unknown, but observers suggest that it can be up to 500 meters. The

GCJ-02 is commonly referred to as **Mars coordinates** because it is as if you're mapping the Earth using WGS 1984 to Mars, or another planet, given the offset that is applied. Commercial mapping applications, such as Google Maps and Baidu Maps, are required to leverage GCJ-02 in their mapping applications within China.

If your analysis pertains to data sourced from within China, know that this will add additional complexities to your work. While there are some studies ongoing related to how to reverse the offset, that work is far from being completed. It is recommended that the answers produced from your analysis when using data from China are caveated as such so that your audience is aware of this condition.

While this is not a comprehensive overview of all of the geographic coordinate systems by any means, it does provide you with an overview of the most common GCS, the WGS 1984 projection. We concluded this section by discussing the GCJ-02 projection and the issues that this presents when working with coordinates from this system. To learn more about the total universe of GCS, we'd recommend that you look at this resource from *ESRI*: `https://desktop.arcgis.com/en/arcmap/latest/map/projections/pdf/geographic_coordinate_systems.pdf`.

Now that you have an understanding of GCS, let's move on to projected coordinate systems.

Understanding projected coordinate systems

The GCS tells you where data is located on the Earth's surface. A PCS tells you how to draw and locate your data on a flat, two-dimensional plane. As mentioned previously, both the Earth and GCS models are spherical. However, most mapping mediums, be it a paper map or a mobile screen, are two-dimensional, flat surfaces. A PCS tells you how to convert the GCS spherical model to a flat model of the Earth's surface.

It is often useful to develop a mental map by thinking about an orange. An orange is typically spherical and can be thought of as the Earth. As you peel an orange, you can lay its peel somewhat flat on a surface, but in order to get it completely flat, you must begin to tear the orange peel. Each tear in the orange peel can be thought of as distortion that is added to the model. A PCS, also known as a **map projection**, is essentially doing the same thing; it is tearing the GCS to allow it to be represented on our flat display mediums. You will want to choose a PCS based on the geographic area you are focused on, such that the distortion is minimized in that area. By minimizing distortion in one area, you are making a trade-off and adding distortion to other areas.

A PCS typically represents locations in linear units, such as meters. This is in contrast to GCS, which represents locations in angular units, such as degrees, as discussed previously.

As with GCSs, there are many different PCSs as well. Each PCS has pros and cons, with some models being better at preserving area-based measurements while others are better at preserving angles. This is another trade-off that you make when identifying the best PCS for your analysis.

Common types of projected coordinate systems

There are a few types of PCS. Each PCS is better or worse at maintaining shape, area, angles, scales, and so on. Knowledge of the trade-offs each map projection makes is critical to ensuring that your analysis is as accurate as possible. If you're working with a specific geography, it is best to select a map projection that is focused on that geography. If you're working on an analysis where comparison of the land area is critical, then you'll need to select a map projection that maintains land area well.

Equal-area projections

Equal-area projections maintain the relative size of an area throughout a map. This means that any given region on a map, and the features of that region, are true to size. In order to maintain area, equal-area projections distort shape and angles and cannot be conformal. *Figure 3.7* shows the continents of Africa and Asia projected in the Lambert cylindrical equal-area projection.

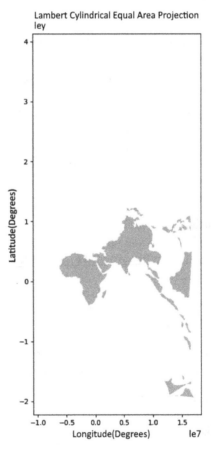

Figure 3.7 – Africa and Asia – Lambert cylindrical equal-area projection

Behrmann map projection

The **Behrmann projection** is a **cylindrical equal-area projection** that was developed by Walter Behrmann in the early 1900s. The standard parallels for this projection fall at 30° N and 30° S. In the Behrmann projection, 50 percent of the Earth's surface is stretched horizontally while the other 50 percent is stretched vertically. This vertical and horizontal stretch helps maintain area at the cost of distorting the shape of the geography. The EPSG code for the Behrmann projection is `54017`.

> **Note**
>
> EPSG codes are maintained by the **European Petroleum Survey Group** (**EPSG**), which organizes geodetic parameters for coordinate systems, datums, and spheroids. The EPSG is an authoritative source for these unique identifiers. Codes maintained by the ESPG are prefaced with the letters ESPG. The other authoritative source is ESRI. ESRI's codes are prefaced with the letters ESRI.

Sinusoidal map projection

The **sinusoidal projection** is a **pseudocylindrical equal-area projection**, which displays all parallels as well as the central meridian at true scale. The meridians that bound the projection bow outwards to an excessive degree, which causes major distortion in shape around the map's outline. The ESPG code for the **MODIS** sinusoidal projection, one of many sinusoidal projections, is **ESPG 6974**. The code for the ESRI World Sinusoidal projection is ESRI `54008`. *Figure 3.8* shows the continents of Africa and Asia projected using the ESRI World Sinusoidal projection.

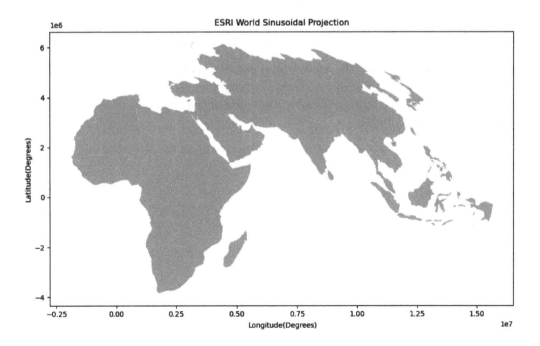

Figure 3.8 – Africa and Asia – ESRI World Sinusoidal Projection

Conformal projections

A **conformal map projection** is one that maintains the mathematical angles between all curves that cross one another on the Earth's surface. This means that if any given roadway crosses another roadway at a 47° angle, then the representation of that on a conformal projection is also 47°. Conformal map projections maintain these angles locally, and thus, added distortion is present as you move away from those local points. In order to maintain angles, the conformal map projection distorts land area by a large margin.

Mercator map projection

The **Mercator map projection** is a **conformal cylindrical map projection**, which means that parallels and meridians cross to form rectangles on the projected map. The Mercator map projection was originally developed in 1569 by Gerardus Mercator. Mercator originally created his namesake map projection to display accurate compass bearing to be used by ships navigating at sea.

In the Mercator projection, the meridians are vertical lines that are parallel and equally spaced to one another, as with a standard GCS. The latitude lines are perpendicular to the parallel longitude lines that begin at the equator. As the latitude lines approach the North and South Poles, they become further and further apart from one another.

The Mercator map projection is very good at maintaining directions, angles, and shapes. However, it does present large distortions in area, in and around the polar regions. When looking at a Mercator map, Greenland appears to be larger than South America. This is, of course, not the case in reality, as Greenland is around one-eighth the size of South America in terms of land area. This distortion makes the Mercator projection most usable for locations that are close to the equator as there is the least amount of distortion in this area. The Mercator projection would not be suitable for comparing areas of towns that are located in South America to those that are located in Greenland due to the distortion.

In modern web and mobile-based GIS, the **Web Mercator** variant of the Mercator projection is the de facto standard. This variant rose to prominence in 2005 when Google Maps adopted it as the standard for their GIS, and it is now used in most other web mapping applications. The EPSG for the Web Mercator projection is 3857.

Figure 3.9 shows the continents of Africa and Asia projected using the Web Mercator Auxiliary Sphere projection.

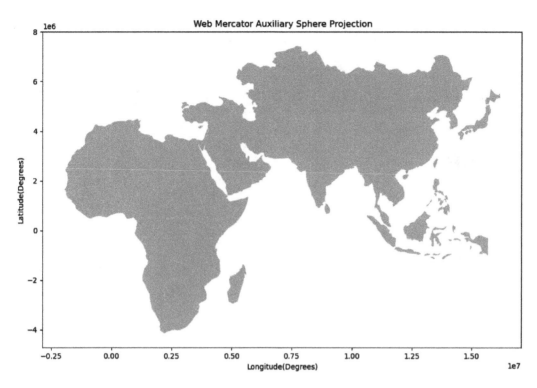

Figure 3.9 – Africa and Asia – Web Mercator Auxiliary Sphere projection

Equidistant projections

Equidistant map projections are map projections that maintain scale along one or multiple lines or from a set of points to all other points on a map.

Equidistant conic map projection

The equidistant conic map projection is a conic map projection that is commonly used in producing maps of smaller countries. The conic map projection is also useful for larger regions and countries whose boundaries stretch east to west across the Earth's surface. Conic map projections have been around for a very long time, as an early version is known as the simple conic projection, which was developed by Ptolemy, a second-century Greek astronomer, and described in his text *Geography*. ESRI maintains the ESRI world equidistant conic projection, which has a standard identifier of ESRI 54027. The North America Equidistant Conic projection is also maintained by ESRI and has an identifier of ESRI 102010. *Figure 3.10* shows the continents of Africa and Asia projected using the ESRI World Equidistant Conic projection.

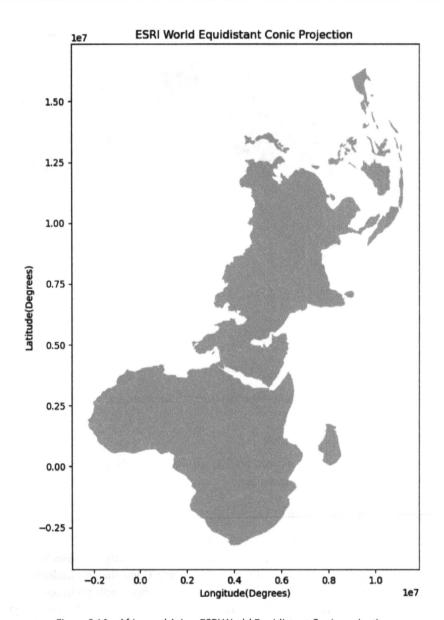

Figure 3.10 – Africa and Asia – ESRI World Equidistant Conic projection

True-direction or azimuthal projections

True-direction projections or **azimuthal projections** maintain the direction from a central point. In order to maintain direction, these projections distort scale. The scale is only true along straight lines radiating out from the center of the map.

Azimuthal equidistant projection

There is a version of the azimuthal projection that maintains both distance and direction, which is known as the **azimuthal equidistant projection**. The World Azimuthal Equidistant projection is maintained by ESRI and has an identifier of ESRI 54032. The North Pole azimuthal equidistant projection is also maintained by ESRI and has an identifier of 102016. *Figure 3.11* shows the continents of Africa and Asia projected using the World Azimuthal Equidistant projection.

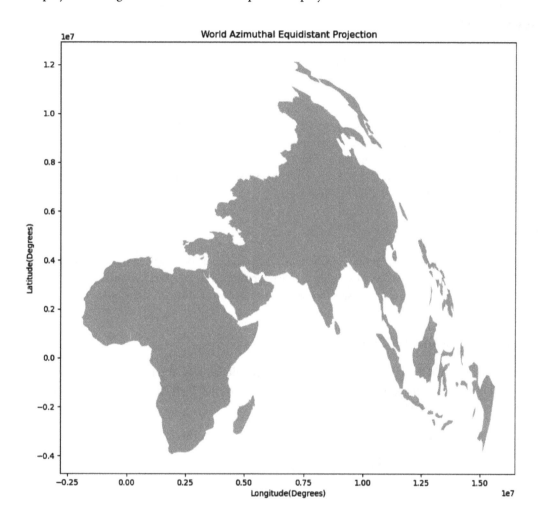

Figure 3.11 – Africa and Asia – World Azimuthal Equidistant projection

In the next section, you'll begin to work hands on with geographic and projected coordinate systems within Python.

Working with GCS and PCS in Python

While we have not yet taken a deep dive into setting up a geospatial data science environment in Python, as that is coming up in *Chapter 4, Exploring Geospatial Data Science Packages*, it is of relevance for us to walk you through how to work with geographic and projected coordinate systems within Python at this time.

There are two notable packages that we'll reference in this section: `PyProj` and `geopandas`. While we won't spend the time in this chapter diving too deep into the weeds of either of these packages, as we'll save that for the next chapter, it is relevant for you to know that both of these packages are useful for working with spatial data and projections.

PyProj

`PyProj` is a Python package used to transform geospatial coordinates from one coordinate reference system into another. The `PyProj` package is useful when working with cartographic projections and geodetic transformations. As of the time of writing this book, the `PyProj` package supports over 100 of the most used map projections. `PyProj` is the backbone projection package that is leveraged within GeoPandas, which we'll discuss briefly next.

GeoPandas

GeoPandas is an extension of Pandas, which adds additional data types for handling spatial data, as well as incremental functions for performing spatial operations. The `geopandas` package also comes with numerous open geospatial datasets that will allow us to work through some quick examples of working with map projections.

Getting hands-on with GeoPandas

In this section, you'll leverage GeoPandas to read in geospatial data. You'll then walk through the following few steps to change the coordinate reference system, and then visualize the data in various projected coordinate systems:

1. To begin, we'll start by importing the requisite packages for this exercise:

    ```
    import geopandas as gpd
    import matplotlib.pyplot as plt
    ```

    It is standard practice to import `geopandas` utilizing the `gpd` shorthand, and `matplotlib.pyplot` using the `plt` shorthand to reference each respective package.

2. We'll then leverage GeoPandas to import three datasets: a `world` shapefile, a `capitals` shapefile, and a `110m` resolution graticule shapefile. Each of these datasets comes from open source data via Natural Earth. To learn more about *Natural Earth* visit https://www.naturalearthdata.com/:

```
world  = gpd.read_file(gpd.datasets.get_
path("naturalearth_lowres"))
capitals = gpd.read_file(data_path + 'ne_110m_populated_
places\\ne_110m_populated_places.shp')
capitals = capitals[capitals["FEATURECLA"]=="Admin-0
capital"]
grat = gpd.read_file(data_path+ 'Graticule\\ne_110m_
graticules_10.shp')
```

3. Next, we need to understand the coordinate reference system that our data is in. To do that, we'll use the `.crs` function from GeoPandas:

```
world.crs
```

This gives us the following output:

```
<Geographic 2D CRS: EPSG:4326>
Name: WGS 84
Axis Info [ellipsoidal]:
- Lat[north]: Geodetic latitude (degree)
- Lon[east]: Geodetic longitude (degree)
Area of Use:
- name: World
- bounds: (-180.0, -90.0, 180.0, 90.0)
Datum: World Geodetic System 1984
- Ellipsoid: WGS 84
- Prime Meridian: Greenwich
```

Here, we can see that the coordinate reference system is `WGS 1984` with `WKID/EPSG 4326`. The prime meridian of this CRS is the Greenwich prime meridian.

4. We now need to ensure that all of our shapefiles are leveraging the same CRS. If the shapefiles are not leveraging the same CRS, this will cause issues in both analysis and mapmaking. To do this, we'll run the following line of code to perform a comparison of the CRS leveraged by each of the shapefiles:

```
capitals.crs == world.crs == grat.crs
```

The output reads `True`, meaning that each of these shapefiles leverages the WGS 1984 CRS.

5. We can now plot a map that overlays the three shapefiles by running the following code:

```
fig, ax = plt.subplots(figsize=(12,10))
world.plot(ax=ax, color="lightgray")
capitals.plot(ax=ax, color="black", markersize=10, marker
="o")
grat.plot(ax=ax, color="lightgray", linewidth=0.5)
ax.set(xlabel="Longitude(Degrees)",
 ylabel="Latitude(Degrees)",
 title="WGS 1984 Datum")
plt.show()
```

This code produces the map visual represented in *Figure 3.12*.

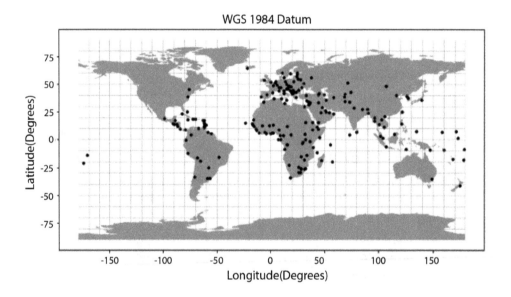

Figure 3.12 – World map with capital cities in CRS WGS 1984

Let's next move on to reprojecting the data.

Reprojecting the data

Now that you've produced a map of the world and capital cities in the standard WGS 1984 CRS, let's play around with reprojecting the data:

1. Let's first reproject the `world` and `graticules` shapefiles into the Azimuthal Equidistant projection we spoke about in the last section. To do that, we'll run the following code:

    ```
    world_ae = world.to_crs("ESRI:54032")
    graticules_ae = grat.to_crs("ESRI:54032")
    ```

2. We'll then run the `.crs` function again to check the metadata of the CRS:

    ```
    world_ae.crs
    ```

    This gives us the following output:

    ```
    <Projected CRS: ESRI:54032>
    Name: World_Azimuthal_Equidistant
    Axis Info [cartesian]:
    - E[east]: Easting (metre)
    - N[north]: Northing (metre)
    Area of Use:
    - name: World
    - bounds: (-180.0, -90.0, 180.0, 90.0)
    Coordinate Operation:
    - name: World_Azimuthal_Equidistant
    - method: Modified Azimuthal Equidistant
    Datum: World Geodetic System 1984
    - Ellipsoid: WGS 84
    - Prime Meridian: Greenwich
    ```

    We can see that the CRS is now labeled `Projected CRS` using `ESRI:54032`, which is the World Azimuthal Equidistant PCS. The prime meridian of this projected coordinate system remains the Greenwich meridian, as in the WGS 1984 CRS.

3. Let's plot the map with these newly projected shapefiles:

    ```
    def plot_map_layers(gdf_1, gdf_2, gdf_3, name,unit):
        fig, ax = plt.subplots(figsize=(12,10))
        gdf_1.plot(ax=ax, color="darkgray")
        gdf_2.plot(ax=ax, color="black", markersize=10,
    marker ="o")
    ```

```
        gdf_3.plot(ax=ax, color="lightgray", linewidth=0.5)
        ax.set(xlabel="X Coordinate -"+ unit,
               ylabel="Y Coordinate -" + unit,
               title=name
               )
plt.show()
plot_map_layers(world_ae, capitals, graticules_ae,
"Azimuthal Equidistant - Unprojected Cities", "Meter")
```

This code produces the map shown in *Figure 3.13*.

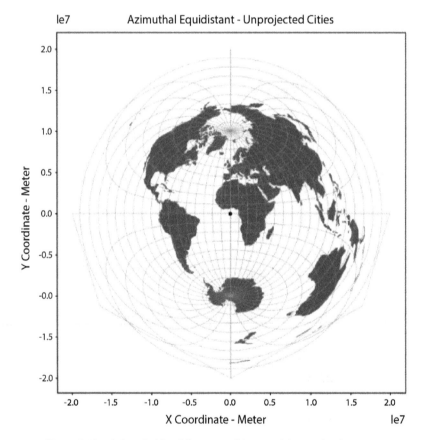

Figure 3.13 – Azimuthal Equidistant world map with a projection error

Wait, what happened to our capital cities layer and why is it showing as a single point?

4. Let's explore this issue by checking the CRS of the capital cities layer

```
capitals.crs
```

Running the prior code block yields the following output:

```
<Geographic 2D CRS: EPSG:4326>
Name: WGS 84
Axis Info [ellipsoidal]:
- Lat[north]: Geodetic latitude (degree)
- Lon[east]: Geodetic longitude (degree)
Area of Use:
- name: World
- bounds: (-180.0, -90.0, 180.0, 90.0)
Datum: World Geodetic System 1984
- Ellipsoid: WGS 84
- Prime Meridian: Greenwich
```

Oh, look at that, the cities layer is still leveraging the WGS 1984 CRS and not the World Azimuthal Equidistant PCS. We must have forgotten to project the capitals shapefile into the new projection.

5. Let's change the projection and then remap our data:

```
capitals_ae = capitals.to_crs("ESRI:54032")
plot_map_layers(world_ae, capitals_ae, graticules_ae,
"Azimuthal Equidistant - Projected Cities", "Meter")
```

With the projection changed in the previous code block, the resulting map now looks like the one depicted in *Figure 3.14*.

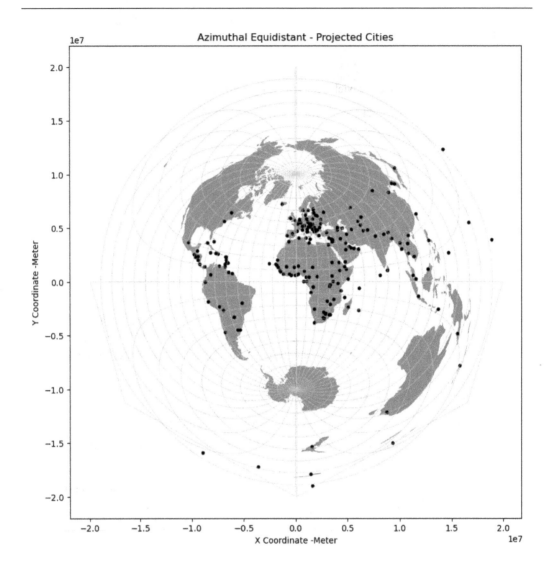

Figure 3.14 – Azimuthal Equidistant world map with capital cities

Now, that is much better. Our capital cities are showing up where we'd expect them to be on the map. While forgetting to change the projection of the capital cities layer was done on purpose to demonstrate the risk of failing to project all data into the same PCS, it shows you how cautious you must be when working with spatial data and coordinate reference systems. Projection errors are one of the most common bugs that can be found in spatial data science workflows.

Summary

We started the chapter off with a history lesson from the ancient Greeks, who began to theorize and test that the Earth was round and not, in fact, flat. With an understanding that the Earth is round, we then dove into GCSs to develop an understanding of how we represent the three-dimensional Earth in two-dimensional space. We talked about various versions of GCSs, such as WGS 1984, which has become the de facto standard GCS, given its high degree of accuracy. We also briefly spoke about the GCJ-02 GCS, which is utilized in China, and the issues it presents due to the randomized offset algorithm.

In the next section, we covered PCSs, which convert angular units, such as degrees used within GCSs, into measurements that use units, such as meters and miles. We covered multiple types of PCSs, including equal-area, conformal, equidistant, and true-direction projections. We also introduced you to the standard identifiers and authoritative sources that maintain projected coordinate systems: ESRI and EPSG.

We concluded the chapter with our first foray into writing Pythonic code for geospatial data science, which leveraged the `geopandas` and `matplotlib.pyplot` packages. During this hands-on exercise, we used data from Natural Earth to plot a map of capital cities using the WGS 1984 CRS. We then projected this data into the World Azimuthal Equidistant PCS. Along the way, we forgot to project our capital cities and learned about the issues that not projecting data correctly can cause in our analysis and mapmaking.

In the next chapter, we'll go deeper into open source geospatial data science Python packages. While we'll leverage packages such as `geopandas` and `matplotlib.pyplot` again, this is just scratching the surface of the total universe of packages you'll leverage throughout your geospatial data science journey!

4

Exploring Geospatial Data Science Packages

Toward the end of *Chapter 3, Working with Geographic and Projected Coordinate Systems*, we introduced you to the Python packages: PyProj, GeoPandas, and Matplotlib. You may recall that we used GeoPandas to read in a shapefile of state capitals, plotted them using Matplotlib, and then projected the capitals using PyProj. Reading in geospatial data, projecting the data, and then plotting it are common steps in numerous geospatial data science initiatives. However, we've just started to scratch the surface in terms of introducing you to the entire universe of geospatial data science packages and the powerful solutions that are just a few keystrokes away.

In this chapter, we'll provide you with a deeper understanding of what PyProj, GeoPandas, and Matplotlib are capable of. We'll also introduce you to a wide array of other geospatial data science packages that you'll rely upon during your work. Some of these packages provide functions for more mainstream applications, while others are utilized for more specialized workflows.

By the end of this chapter, you'll have an understanding of the following:

- Python packages that enable reading and writing of vector and raster data
- Python packages enabling spatial analysis and modeling
- Python packages that assist in producing high-quality mapping visualizations

Technical requirements

For this chapter, there are several Jupyter notebooks that are within the book's GitHub repository. Each notebook begins with `Chapter 4 -`, followed by the name of the package that is to be discussed. The book's GitHub can be accessed at the following link: `https://github.com/PacktPublishing/Applied-Geospatial-Data-Science-with-Python/tree/main/Chapter04`.

Packages for working with geospatial data

There are many packages that enable you to work with geospatial data in Python. In this section, we'll discuss some of the most common packages that you'll interact with during the course of common geospatial data science workflows.

GeoPandas

As we mentioned in the prior chapter, GeoPandas is an extension of pandas, which adds support for additional data types necessary for working with spatial data. It also includes additional methods not found in pandas, which enable you to perform spatial operations and produce spatial data visualizations. We'll discuss pandas later on in this chapter in the *Reviewing foundational data science packages* section. pandas is a foundational package required for most general data science workflows.

The core functionality of GeoPandas includes the following:

- Reading and writing spatial data
- Spatial data structures
- Projection management
- Spatial data visualization
- Data manipulation
- Geocoding

Let us now go into more detail on each of these core functions.

Reading and writing spatial data

GeoPandas natively supports reading and writing almost any vector-based dataset. This includes the ESRI shapefile and GeoJSON file formats, which we introduced you to in *Chapter 2, What Is Geospatial Data and Where Can I Find It?* In order to read in files, GeoPandas relies on the Python package **Fiona**. We'll discuss Fiona further later on in this section.

Reading and writing spatial data

To read a vector file into GeoPandas, you'll need to leverage the following command:

```
geopandas.read_file()
```

As we mentioned in the previous chapter, you can import the GeoPandas library using its standard shorthand, or alias, gpd:

```
import geopandas as gpd
```

A brief aside on package imports and their alias names

When you begin a new data science workflow, it is necessary to import the packages or libraries that are relevant to the forthcoming analysis. When we talk about importing packages, we are telling Python to bring in a previously installed package into the working environment where we're performing our analysis.

Importing libraries can be done using the standard import for that library, such as `import geopandas`. It can also be done using the shorthand, or alias version, which looks like `import geopandas as gpd`. Leveraging an alias reduces the amount of typing that you'll have to do throughout your workflow. As you grow as a geospatial data scientist, you'll learn the common aliases of most Python packages. For those packages we discuss in this chapter, we'll provide you with their standard alias.

Inside the parenthesis of the `read_file` method, you'll begin passing parameters. The first parameter passed to `read_file` will be the file path of the data you're reading in. Looking back at the code we leveraged in the last chapter, we wrote the following to read in the capital's data:

```
capitals = gpd.read_file(data_path + 'Graticule\\ne_110m_
graticules_10.shp')
```

In this code, we're using the `read_file` method to read in `capitals` from the ne_110m_graticules_10.shp file, stored in the data folder on our local machine. GeoPandas, because it relies on Fiona, is relatively smart in its association of file type and the file type driver necessary to read in that file. So, we don't have to specifically tell GeoPandas that we are reading in a shapefile. However, if we run into an error, we can specify the driver parameter and set it to `Shapefile` or another driver:

```
capitals = gpd.read_file(data_path + 'Graticule\\ne_110m_
graticules_10.shp', driver='shapefile')
```

The `GeoPandas.read_file` method doesn't require that the file being read in be stored locally on your machine or even a flat file. To read in files from a URL path, you can leverage the following code:

```
# Reading from a URL
url = "https://d2ad6b4ur7yvpq.cloudfront.net/
naturalearth-3.3.0/ne_110m_admin_1_states_provinces_shp.
geojson"
us_states = gpd.read_file(url)
```

To read in files from a zipped file, you can use code similar to this:

```
# Reading in data stored in a zipped file
us_cbsas = gpd.read_file(data_path + 'tl_2021_us_cbsa.zip')
```

In version 0.7.0 of GeoPandas, the developers added the ability to filter an input dataset based on the geometry of another object. To demonstrate this, we'll write and execute code that will read in the US state of California and select all of the US Census Bureau **core-based statistical areas** (**CBSAs**) that intersect with California:

```
# Read in the US States file from the Census Tiger Files saved
as a zip
us_states = gpd.read_file(data_path + 'tl_2021_us_state.zip')
# Filter the US States file to be just California
california = us_states[us_states['NAME']=="California"]
# Show the dataframe
california.head()
# Create a new geopandas dataframe that only includes the CBSAs
that are in California using the geopandas mask parameter and
passing the previously defined california geopandas dataframe
ca_cbsas = gpd.read_file(data_path + 'tl_2021_us_cbsa.zip',
mask=california)
# Show the first 5 records of the ca_cbsas geopandas dataframe
ca_cbsas.head()
```

The California GeoPandas dataframe is displayed in *Figure 4.1*.

	REGION	DIVISION	STATEFP	STATENS	GEOID	STUSPS	NAME	LSAD	MTFCC	FUNCSTAT	ALAND	AWATER	INTPTLAT	INTPTLO
13	4	9	06	01779778	06	CA	California	00	G4000	A	403671756816	20293573058	+37.1551773	-119.543418

Figure 4.1 – California GeoPandas dataframe

The first five rows of the resulting dataframe, after using the California GeoPandas dataframe as a mask, is displayed in *Figure 4.2*.

	CSAFP	CBSAFP	GEOID	NAME	NAMELSAD	LSAD	MEMI	MTFCC	ALAND	AWATER	INTPTLAT	INTPTLON	geometry
0	None	12540	12540	Bakersfield, CA	Bakersfield, CA Metro Area	M1	1	G3110	21068632654	78405068	+35.3466288	-118.7295064	POLYGON ((-118.88168 34.81785, -118.88204 34.8...
1	None	15060	15060	Brookings, OR	Brookings, OR Micro Area	M2	2	G3110	4217489863	934614482	+42.4664387	-124.2109292	POLYGON ((-123.82239 42.15933, -123.82246 42.1...
2	None	17340	17340	Clearlake, CA	Clearlake, CA Micro Area	M2	2	G3110	3254452700	188748214	+39.0948019	-122.7467569	POLYGON ((-122.41579 38.76800, -122.41584 38.7...
3	None	18860	18860	Crescent City, CA	Crescent City, CA Micro Area	M2	2	G3110	2606118035	578742633	+41.7499033	-123.9809983	POLYGON ((-124.31611 41.72839, -124.33061 41.7...
4	456	22280	22280	Fernley, NV	Fernley, NV Micro Area	M2	2	G3110	5187939862	59489121	+39.0222125	-119.1974246	POLYGON ((-118.90650 38.51631, -118.90649 38.5...

Figure 4.2 – California CBSAs GeoPandas dataframe

Another incremental improvement came with GeoPandas version 0.1.0 when developers added an additional filtering mechanism using **polygonal bounding boxes**. A bounding box is a polygon that surrounds an object. Here, you'll use a bounding box to filter the `ca_cbsas` dataframe.

To filter a GeoPandas dataframe using a bounding box, you'll first need to define the bounding box, and then pass that to the `bbox` parameter of the `read_file` method:

```
# Defining the bounding box
bounding_box = (-123.82239, 42.15933, -123.82246, 38.7)

#Reading in the CA_CBSAs and filtering based on the bounding
box
cbsas_bbox = gpd.read_file(zipped_file, bbox=bounding_box)

# Showing the first 5 rows of the filtered dataframe
cbsas_bbox.head()
```

To write spatial data with GeoPandas, you need to leverage the `to_file` method. The `to_file` method works very similarly to the `read_file` method, except you are now creating a new file on your local machine instead of reading from an existing file. Let us begin by writing out the previously defined `ca_cbsas` file as a shapefile. To do this, you'll define an output file path and then run the `to_file` method on the `ca_cbsas` GeoPandas dataframe. The code to perform this operation is as follows:

```
# Setting the folder we want to write the output data to
out_path = r"YOUR FILE PATH"

# Writing out the data as a shape file
ca_cbsas.to_file(out_path+"ca_cbsas.shp")
```

Similar to the `read_file` method, you can also work with other file types with the `to_file` method. In the following code, we'll write out the dataframe that was filtered using the bounding box as a GeoJSON file. Here you'll need to specify the driver as GeoJSON, as it is not a default driver like shapefile:

```
# Writing out the data as a shape file
cbsas_bbox.to_file(out_path+"cbsas_bbox.geojson",
driver="GeoJSON")
```

In this section, we've started to introduce you to the GeoPandas dataframe, one of the data structures present in GeoPandas. We'll dive deeper into the topic of data structures in the next section.

Spatial data structures

In the previous section, we mentioned the GeoPandas dataframe quite often. Before we go too much further, it's best we take a minute to discuss the spatial data structures that you'll work with when using GeoPandas. The GeoPandas dataframe, also known as the **GeoDataFrame**, is one such structure. The other main data structure within GeoPandas is the **GeoSeries**. Both GeoSeries and GeoDataFrame are subclasses of the pandas Series and dataframe data structures.

The GeoSeries data structure can be thought of as a vector. Each element, or row, of the vector is a shape that corresponds to one observation or record. Each entry in a vector can be a single shape, such as a point, line, or polygon. The entry can also be a more complicated polygon that could represent an entity, such as all of the states in the United States as one record in a dataset of countries and their component regions. GeoPandas supports the three vector structures we introduced you to previously, which are points, lines, and polygons. GeoPandas also supports multi-point, multi-line, and multi-polygon geometric objects.

Let us now review some of the attributes and methods of the GeoSeries data structure:

- Attributes:

  - `area()`: This provides the area of the geometry in the units defined by the spatial projection.

  - `geom_type()`: This provides the type of geometry for each element of the GeoSeries.

  - `bounds()`: This provides a tuple of the minimum and maximum coordinates for both axis for each element.

  - `total_bounds()`: This provides a tuple of the minimum and maximum coordinates across each axis across the GeoSeries instead of by element.

- Methods:

  - `to_crs()`: This changes the coordinate reference system of the data. We introduced you to this method in *Chapter 3, Working with Geographic and Projected Coordinate Systems*.

  - `distance()`: This returns a series with the minimum distance from each entry to other entries.

  - `centroid()`: This returns the **centroid** of the elements in the GeoSeries. A centroid is the geometric center of a geometric object.

The GeoPandas GeoSeries can also perform a test based on spatial relationships. A **spatial relationship** is defined as the way a set of two or more objects are set in relation to one another in geographic space. Another term for spatial relationships is **spatial topology**. Spatial relationships are typically based on adjacency, contiguity, overlap, and proximity. Two main spatial relationships included within the GeoSeries are `intersects()` and `contains()`. `intersects()` checks whether two objects intersect or overlap one another. The `contains()` test measures whether a shape is contained within another shape.

Now that we've introduced you to the GeoSeries data structure, let's talk about the GeoDataFrame structure. The GeoDataFrame is essentially a collection of GeoSeries and can be thought of as similar to any tabular data structure you may have worked with in the past. The most important and unique feature of a GeoDataFrame is that it has a special GeoSeries that is referred to as a GeoDataFrame's geometry. The geometry GeoSeries stores the geometry of each element in the GeoDataFrame. When a method or spatial relationship test is performed on a GeoDataFrame, it relies on the geometry GeoSeries. To access the geometry GeoSeries, you can use the `.geometry` attribute, regardless of whether the column is named geometry or renamed something else. Each cell in the geometry GeoSeries is a Shapely geometry which comes from the Shapely package. We'll discuss Shapely in more depth later on in this chapter.

In the next section, we'll discuss how to manage spatial data projections within GeoPandas.

Projection management

As we discussed at the end of *Chapter 3, Working with Geographic and Projected Coordinate Systems*, GeoPandas can be used to manage the projection of geospatial data. To change the projection of a GeoPandas data structure, you use the `to_crs()` method on the data structure. The `to_crs()` method in GeoPandas functions by making a subcall to the `PyProj` package. The specific method within PyProj that is being called is the `PyProj.crs.from_user_input()` method. Thus, the GeoPandas `to_crs()` method will accept anything that is accepted by the `PyProj` method, including, but not limited to, the following:

- A CRS **well-known text (WKT)** string
- An ESPG numeric identifier, such as `4326`, or an authority string, such as `espg:4326`
- A `PyProj.crs` class

Now that you understand how to read in spatial data and manipulate its projection, let's talk about how you can visualize geospatial data using GeoPandas.

Spatial data visualization

In *Chapter 3, Working with Geographic and Projected Coordinate Systems*, we leveraged a package called Matplotlib to produce some very basic visualizations of capital cities across the world. Within GeoPandas, there exist a set of methods that provide a high-level interface with Matplotlib. To do this, you can simply use the `.plot()` method on a GeoSeries or a GeoDataFrame object.

As an example, we can plot the `ca_cbsas` object that we defined earlier with one simple line of code:

```
ca_cbsas.plot()
```

The output of this code is represented in *Figure 4.3*.

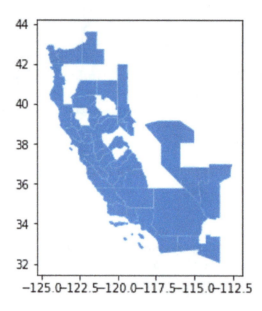

Figure 4.3 – California CBSAs map

The high-level mapping functionality of GeoPandas also makes it quick and easy to produce **choropleth maps**. A choropleth map is a map where each geometry on the map is colored based on a value associated with the geometry. This could include things such as population density or the amount of tax revenue in each geography. To produce a choropleth map, we need to read in some new data. Luckily, GeoPandas comes preloaded with some basic data. For this example, you'll leverage the `naturalearth_lowres` data to produce a choropleth map of the population of each country:

```
# Choropleth Map of the world's population
# Reading in the natural earth lowres data
world = gpd.read_file(gpd.datasets.get_path('naturalearth_
lowres'))


# Plotting the data colored by the pop_est GeoSeries
world.plot(column='pop_est')
```

The result of running this code is the choropleth map shown in *Figure 4.4*.

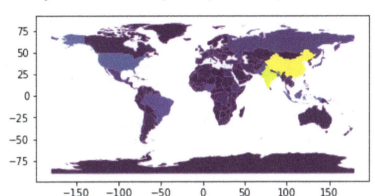

Figure 4.4 – A choropleth map showing the country's population

As of right now, this map isn't very good. Not only is the map small, but it also lacks important facets of a good map, such as axis titles and a legend that will provide the end user with the necessary context to interpret it. For now, we'll skip over fixing these issues and will come back to it later on in *Chapter 5, Exploratory Data Visualization*.

Data manipulations

There are numerous data and geometric manipulations that can be performed with GeoPandas. While there are too many manipulations to cover at length in this chapter, we will cover the ones that you will most frequently interact with in your day-to-day work.

Data manipulations include the following options:

- **Append**: The GeoPandas `.append()` method makes a call to the pandas `.append()` method. The `.append()` method is used to append data from two objects at a row level. This means that the information from object two is added at the end of object one. It is important to keep in mind that the data to be appended must have the same CRS.

- **Join**: To join two GeoDataFrames together based on a common identifier, it is best to use the `.merge()` method. This method takes two parameters by default, the GeoDataFrame object to be merged and a common identifier passed to the `on` parameter. By default, the `.merge()` method performs a left join. In order to perform other types of joins, you'll need to pass information to the `how` parameter, which can take the following inputs: `left`, `right`, `outer`, `inner`, and `cross`.

Geometric manipulations include the following options:

- **Buffer**: The `.buffer()` method takes a distance input passed to the `distance` parameter and returns a GeoSeries of geometries, which represents the points that fall within the given distance of the geometric object.

- **Centroid**: The `.centroid()` method returns a GeoSeries representing the centroid of the geometric object.

- **Simplify**: The `.simplify()` method returns a GeoSeries that represents a simplified, or smoothed, version of the input geometries. This helps reduce the memory needed by reducing the detail of complex geometries.

- **Spatial join**: A spatial join combines two GeoDataFrames based on the spatial relationship of the observations within those two GeoDataFrames. To do this, you'll need to leverage one of two spatial join methods in GeoPandas, the `.sjoin()` method or the `.sjoin_nearest()` method. The `.sjoin()` method joins based on binary spatial relationships, such as `intersect` or `contains`, while the `.sjoin_nearest()` method joins data based on the proximity of the observations.

- **Dissolve**: The `.dissolve()` method is used on a lower-order geography in order to create a higher-order geography. Take, for example, the GeoDataFrame in *Figure 4.5*, where each observation represents a country in the world and, for your analysis, you need data at a continent level.

	pop_est	continent	name	iso_a3	gdp_md_est	geometry
0	920938	Oceania	Fiji	FJI	8374.0	MULTIPOLYGON (((180.00000 -16.06713, 180.00000...
1	53950935	Africa	Tanzania	TZA	150600.0	POLYGON ((33.90371 -0.95000, 34.07262 -1.05982...
2	603253	Africa	W. Sahara	ESH	906.5	POLYGON ((-8.66559 27.65643, -8.66512 27.58948...
3	35623680	North America	Canada	CAN	1674000.0	MULTIPOLYGON (((-122.84000 49.00000, -122.9742...
4	326625791	North America	United States of America	USA	18560000.0	MULTIPOLYGON (((-122.84000 49.00000, -120.0000...

Figure 4.5 – World GeoDataFrame

To convert country data to continent-level data, you can run the `.dissolve()` method on the world GeoDataFrame and pass the column `continent` to the method's by parameter. This returns the continents' GeoDataFrame, represented in *Figure 4.6*.

continent	geometry
Africa	MULTIPOLYGON (((32.830 -26.742, 32.580 -27.470...
Antarctica	MULTIPOLYGON (((-163.713 -78.596, -163.713 -78...
Asia	MULTIPOLYGON (((120.295 -10.259, 118.968 -9.55...
Europe	MULTIPOLYGON (((-51.658 4.156, -52.249 3.241, ...
North America	MULTIPOLYGON (((-61.680 10.760, -61.105 10.890...

Figure 4.6 – continents GeoDataFrame

Plotting the newly defined `continents` dataframe produces the map in *Figure 4.7*.

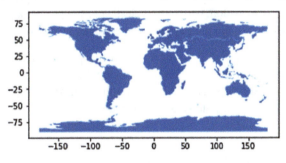

Figure 4.7 – continents plot

The code needed to perform these operations is as follows:

```
world.head()
continents = world.dissolve(by="continent")
continents.head()
continents.plot()
```

This concludes our section on GeoPandas' data manipulations. In the next section, we'll talk about one final and very important functionality of GeoPandas, geocoding.

Geocoding

In addition to reading in data that already includes geographic attributes, you can also create new geographic data from addresses using GeoPandas. The process of converting addresses or place names to geographic coordinates is called **geocoding**. To perform a geocoding exercise in GeoPandas, you need to leverage the GeoPy package, which is an optional dependency of GeoPandas. The GeoPy package takes as its input a pandas DataFrame of addresses or locations without a geometry attribute column and returns a GeoDataFrame with the input address and the resulting geometry column.

To perform a geocoding operation, you'll first need to import `geocode` from the `geopandas.tools` package. The `geocode` function takes multiple inputs, with the first being a `pandas series` that holds the addresses to be geocoded. The second required parameter is `provider`, which GeoPy is leveraging to perform the geocoding operation. GeoPandas and GeoPy, in and of themselves, are not geocoding services, and thus, the packages rely on the geocoding services provided by third parties. After you pass the provider information, you'll also typically need to provide an **API key** that securely connects you to that provider through their **application programming interface** or **API**.

Let's see how this works in practice. You'll start by importing the pandas package so that you can read in a CSV file, which contains a list of addresses to three popular Washington, DC, attractions: the Washington Monument, the Smithsonian National Air and Space Museum, and the White House. To do that, you'll need to execute the following lines of code:

```
import pandas as pd
dc_attractions = pd.read_csv(data_path + 'DC_Attractions.csv')
dc_attractions.head()
```

This yields the DataFrame displayed in *Figure 4.8* with the name of the attraction and its corresponding address.

Attraction	Address
Washington Monument	2 15th St NW, Washington, DC 20024
Smithsonian National Air and Space Museum	600 Independence Ave SW, Washington, DC 20560
White House	1600 Pennsylvania Avenue NW, Washington, DC 20500

Figure 4.8 – Washington, DC, attractions

You can then import the `geocode` function and pass to it the address column of the pandas DataFrame.

```
from geopandas.tools import geocode
dc_attractions_gpd = geocode(dc_attractions['Address'],
provider='openmapquest', api_key="APIKEY")
dc_attractions_gpd.head()
```

This yields the GeoPandas GeoDataFrame displayed in *Figure 4.9* with the address and the geometry output from the geocoding operation.

	geometry	address
0	POINT (-77.03459 38.90178)	15th Street NW, Golden Triangle, Washington, D...
1	POINT (-77.03771 38.88772)	Independence Avenue Southwest, Penn Quarter, W...
2	POINT (-77.03655 38.89772)	1600, Pennsylvania Avenue Northwest, Golden Tr...

Figure 4.9 – Geocoded Washington, DC, attractions

In the previous line of code, we've chosen to use the OpenMapQuest service, which is provided by MapQuest. To leverage this same service, you'll need to visit `https://developer.mapquest.com/user/login/sign-up`. Once there, fill out their free sign-up form and then visit the **Manage Keys** tab to create your API key for this application. You'll then need to paste that API key into the quotes containing `APIKEY` in the code.

GDAL

GDAL stands for the **Geospatial Data Abstraction Library**, and it is made up of two components, GDAL, which is used for manipulating and working with raster data, and OGR, which is used when working with vector data files. During your time as a geospatial data scientist, you likely won't spend a ton of time directly interacting with GDAL, but it is important to know that GDAL provides most of the backend tools, or bindings, that are necessary for working with geospatial data in Python. These bindings are Pythonic wrappers around C++ code. To learn more about GDAL, we recommend you visit the GDAL documentation site at: `https://gdal.org/api/python.html`.

Shapely

We briefly introduced you to the Shapely package earlier on in this chapter in our discussion on GeoPandas. Shapely is a Python package that is used when working with vector geometries and it includes functions that allow you to create geometries and others that allow you to perform operations on those geometries. At its root, Shapely is a Python interface for **Geometry Engine Open Source software** or **GEOS**. GEOS software is embedded in other open source software, including QGIS.

In contrast to GeoPandas, Shapely is only suited for working with one individual geometry at a time, and also, it does not possess any functions for reading or writing geospatial data. To perform functions on multiple geometries or to read and write geospatial data, you'll need to leverage GeoPandas, which is built on top of Shapely.

At times, you may need to interact with Shapely to create geometric data. To do that, you'll need to leverage the geometry and wkt subpackages of Shapely. You'll do this through the following code:

```
# importing shapely subpackages
import shapely.geometry
import shapely.wkt
```

The shapely.wkt subpackage is used to convert a WKT string to a Shapely geometry. You may recall that we briefly introduced you to WKT in *Chapter 2, What Is Geospatial Data and Where Can I Find It?* To elaborate further, WKT is a text markup language that is used to represent vector geometries. To convert WKT to a Shapely geometry, we'll pass the string to the shapely.wkt. loads function. In the following example code, we'll create an object called s_poly1 and display the polygon by directly calling the object:

```
s_poly1 = shapely.wkt.loads("POLYGON ((0 0, 0 -2, 9 -2, 9 0, 0
0))")
s_poly1
```

The displayed polygon is shown in *Figure 4.10*.

Figure 4.10 – WKT derived Shapely geometry

You'll notice that the first and last coordinates in the code that produced *Figure 4.10* are the same. This is necessary to close the polygon, which then becomes a long rectangle.

> **Note**
>
> If you don't repeat the same coordinate to close the polygon, an error message will present itself stating: WKTReadingError: Could not create geometry because of errors while reading input. Another error message stating: IllegalArgumentException: Points of LinearRing do not form a closed linestring is also produced.

You can also produce the same polygon by passing data formatted as GeoJSON. This is shown in the next code block:

```
# Creating the same polygon from GeoJSON
from shapely.geometry import Polygon
geo = {'type': 'Polygon',
  'coordinates': [[[0, 0],
```

```
   [0, -2],
   [9, -2],
   [9, 0]]]}
Polygon([tuple(l) for l in geo['coordinates'][0]])
```

Shapely can also produce multipolygons from WKT, which is demonstrated by the following code:

```
# Shapely Multi-Polygon
s_poly3 = shapely.wkt.loads("""
MULTIPOLYGON
(((50 40, 20 45, 45 30, 50 40)),
((25 35, 10 30, 10 18, 30 5, 45 20, 25 35), (32 20, 26 15, 20
25, 32 20)))
""")
s_poly3
```

The result of executing the code is the multi-part polygon that is displayed in *Figure 4.11*.

Figure 4.11 – WKT derived Shapely multipolygon

In addition to converting from WKT, you can also create Shapely geometries by passing a list of coordinates to unique functions to produce your desired geometry. The list of Shapely functions that produce specific geometries is described in *Table 4.1*.

Shapely geometry	Function
Polygon	`shapely.geometry.Polygon()`
MultiPolygon	`shapely.geometry.MultiPolygon()`
Point	`shapely.geometry.Point()`
MultiPoint	`shapely.geometry.MultiPoint()`
LineString	`shapely.geometry.LineString()`
MultiLineString	`shapely.geometry.MultiLineString()`
GeometryCollection	`shapely.geometry.GeometryCollection()`

Table 4.1 – Shapely geometry functions

As an example of one of these functions in action, let's leverage the `shapely.geometry.LineString()` function to make a line:

```
# Produce a Shapely LineString geometry from a list of
coordinates
coords = [(5, 0.5), (5, 3), (-2, 0), (8, 0)]
line = shapely.geometry.LineString(coords)
line
```

The result is the line displayed in *Figure 4.12*.

Figure 4.12 – Shapely LineString geometry

Now that we've produced a few Shapely geometries, let's talk about the derived properties of Shapely geometries. The three main derived properties are bounds, length, and area. These properties are discussed in the following *Table 4.2*.

Derived Property	Functionality
`.bounds`	Produces a tuple that represents the upper-right, upper-left, bottom-left, and bottom-right coordinates representing the outer bounds of the geometry.
`.length`	Calculates the length of the geometry. For Point and MultiPoint geometries, the length is 0, and for Polygon and MultiPolygon geometries, the length is the sum of the perimeters of the exteriors and any interiors of the polygon.
`.area`	Calculates the area of the geometry. The area is non-zero for Polygons and MultiPolygon geometries.

Table 4.2 – Shapely derived properties

Shapely geometries can also be used to derive new geometries by using some common methods, which are described in the following *Table 4.3*.

Method	Functionality
`.buffer()`	Returns a buffer around the geometry
`.centroid`	Returns the centroid of the geometry
`.envelope`	Returns a bounding box around the geometry
`.convex_hull`	Returns the minimal convex polygon that surrounds a given set of points
`.simplify`	Returns a simplified geometry from the more detailed geometry
`.intersection`	Returns the shared areas of two or more input geometries
`.union`	Returns a geometry representing the total geometry covered by two or more input geometries
`.difference`	Returns the geometry covered by one geometry but not covered by another

Table 4.3 – Shapely derived geometry methods

The final topic we'll cover on Shapely is its Boolean operations, which evaluate the relationships between geometries. The method is run on one Shapely geometry object with a second geometry object passed to the method. Those Boolean operations are described in *Table 4.4*:

Method	Functionality
`.equals`	Evaluates whether the geometry of the two objects is equal
`.almost_equals`	Evaluates whether the geometry of the two objects is approximately equal
`.covers`	Returns true if all points of the other geometry are within the geometry of the object the method is run against
`.crosses`	Returns true if the geometries of the two objects cross at any point
`.contains`	Evaluates whether the geometry object the method is run against contains the geometry of the other object
`.covered_by`	Returns true if the other geometry covers the geometry of the object the method is run against
`.overlaps`	Returns true if the geometries have more than one but not all points in common
`.intersects`	Returns true if the geometries intersect in any way with regard to their boundaries or interiors

Table 4.4 – Shapely Boolean operations

Additional information on Shapely can be found in the Shapely User Manual found at `https://shapely.readthedocs.io/en/stable/manual.html`. This concludes our overview of Shapely. Next, we'll talk about Fiona.

Fiona

Fiona is a simple and easy-to-understand Python package used for reading and writing geospatial data. Fiona leverages standard Python types and protocols, including files, mappings, dictionaries, and iterators, instead of classes that are specific to OGR. Fiona reads geospatial data in a way that is modeled on the GeoJSON file format. Fiona also easily integrates with other geospatial packages, including pyproj, Rtree, and Shapely.

To read and write data with Fiona, you'll leverage the `fiona.open()` function, which returns a file-like `Collection` object. To do this, you'll simply import Fiona and then open the file containing the geospatial data. The following example reads in the CBSA file that you worked with previously:

```
# Importing Fiona
import Fiona
# Reading in the CBSA file
f_obj = fiona.open(data_path + 'ca_cbsas.shp', 'r')
f_obj
```

This code returns the `Collection` object, shown in *Figure 4.13*.

```
<open Collection 'G:\My Drive\Geospatial Data Science with Python\Data\ca_cbsas.shp:ca_cbsas', mode 'r' at 0x216a3f165c8>
```

Figure 4.13 – The CBSA Collection object

The default mode for Fiona is `'r'`, which stands for read. To write a geospatial file, you'll need to change this mode to `'w'`, which stands for write.

Fiona's collection data structure supports many of the same geometry types that are included in GeoPandas and Shapely, including Point, LineString, Polygon, MultiPoint, MultiLineString, and MultiPolygon. Fiona also supports similar methods to those that are included with GeoPandas. For instance, running the `.crs` method on a Fiona collection returns the coordinate reference system of that collection.

Rasterio

Up until this point, we've only discussed Python packages that are useful when working with vector data. As you may recall from *Chapter 2*, vector data is only one part of the geospatial data ecosystem, with raster data being the other component. In this section, we'll be discussing the Python package `rasterio`, which is specially developed to work with raster data formats. Rasterio is built on top of and is compatible with the NumPy Python package, which we'll introduce you to later in this chapter in the *Reviewing foundational data science packages* section.

. Rasterio enables you to do the following:

- Read and write raster files

- Examine the properties of raster files

- Convert a NumPy array into a raster file

- Perform calculations with raster files

- Visualize raster files

In comparison to GeoPandas' ability to work with vector data and perform numerous functions upon that data, the Rasterio package is not as comprehensive and requires other packages to extend its functionality.

To begin this section, you'll need to import a handful of packages, which will be leveraged throughout the section. To do that, you'll need to execute the following lines of code:

```
# Importing the required Python packages
import rasterio
from rasterio.plot import show
import numpy as np
import shapely.geometry
import geopandas as gpd
import glob
```

Processing raster data with Rasterio

To begin reading in raster file formats, such as the GeoTiff format we introduced you to in *Chapter 2*, you'll need to leverage the `rasterio.open` function. The `rasterio.open` function makes a connection to the raster data and immediately reads in the properties of it. Rasterio waits until the data is needed for a downstream function or process to read in the actual data as a way to save the memory overhead that is required to read in large raster files.

In this section, you'll work with data from the **United States Forest Service (USFS) Landscape Change Monitoring System (LCMS)**. The particular data set you'll work with is the annual land cover data for the year 2021. To read in the file, you'll need to pass the file path and `mode='r'` to `rasterio.open`, as shown in the next code block:

```
lc = rasterio.open(data_path + "LCMS_CONUS_v2021-7_Land_Cover_
Annual_2021\\LCMS_CONUS_v2021-7_Land_Cover_2021.tif", mode='r')
lc
```

You can then call the `lc` object, which is displayed in *Figure 4.14*, to view the connection.

```
<open DatasetReader name='G:/My Drive/Geospatial Data Science with Python/D
ata/LCMS_CONUS_v2021-7_Land_Cover_Annual_2021/LCMS_CONUS_v2021-7_Land_Cover
_2021.tif' mode='r'>
```

Figure 4.14 – Rasterio connection to 2021 land cover data

The file path will be different based on where you downloaded and stored the data folder.

> **Note: The Rasterio connection**
>
> The connection that is made to the GeoTiff raster data using Rasterio is similar to the text connection string that the Fiona package makes to vector data. You'll also notice that the `.open()` functions of both packages require `'r'`, standing for read, and `'w'`, standing for write, to be passed to the function in order to differentiate between reading and writing data.

With a connection now established to the raster file, you can now review some of the properties of the raster data file. The `.name` method displays the name of the raster file while the `.meta` method displays the **metadata**; that is data about the data regarding the raster file:

```
lc.name
```

Executing the prior line of code yields the information on the `lc` object that is displayed in *Figure 4.15*.

```
'G:/My Drive/Geospatial Data Science with Python/Data/LCMS_CONUS_v2021-7_La
nd_Cover_Annual_2021/LCMS_CONUS_v2021-7_Land_Cover_2021.tif'
```

Figure 4.15 – Raster name

The next code block calls the `.meta` parameter to understand the metadata associated with the file:

```
lc.meta
```

Calling the `.meta` function as was done in the prior code cell results in an output of all the metadata for the `lc` object. This information is displayed in *Figure 4.16*.

```
{'driver': 'GTiff',
 'dtype': 'uint8',
 'nodata': 0.0,
 'width': 154180,
 'height': 97279,
 'count': 1,
 'crs': CRS.from_wkt('PROJCS["Albers_Conical_Equal_Area",GEOGCS["WGS 84",DA
TUM["WGS_1984",SPHEROID["WGS 84",6378137,298.257223563,AUTHORITY["EPSG","70
30"]],AUTHORITY["EPSG","6326"]],PRIMEM["Greenwich",0],UNIT["degree",0.01745
32925199433,AUTHORITY["EPSG","9122"]],AUTHORITY["EPSG","4326"]],PROJECTION
["Albers_Conic_Equal_Area"],PARAMETER["latitude_of_center",23],PARAMETER["l
ongitude_of_center",-96],PARAMETER["standard_parallel_1",29.5],PARAMETER["s
tandard_parallel_2",45.5],PARAMETER["false_easting",0],PARAMETER["false_nor
thing",0],UNIT["metre",1,AUTHORITY["EPSG","9001"]],AXIS["Easting",EAST],AXI
S["Northing",NORTH]]'),
 'transform': Affine(30.0, 0.0, -2361585.0,
        0.0, -30.0, 3177435.0)}
```

Figure 4.16 – Raster metadata

The metadata provides us with useful information, such as the driver used to read the raster file, the CRS, and the count of records in the raster file. Each of these pieces of data can be directly accessed using the following methods: .driver, .crs, and .count. The output of calling the .crs method is displayed in *Figure 4.17*.

```
CRS.from_wkt('PROJCS["Albers_Conical_Equal_Area",GEOGCS["WGS 84",DATUM["WGS
_1984",SPHEROID["WGS 84",6378137,298.257223563,AUTHORITY["EPSG","7030"]],AU
THORITY["EPSG","6326"]],PRIMEM["Greenwich",0],UNIT["degree",0.0174532925199
433,AUTHORITY["EPSG","9122"]],AUTHORITY["EPSG","4326"]],PROJECTION["Albers_
Conic_Equal_Area"],PARAMETER["latitude_of_center",23],PARAMETER["longitude_
of_center",-96],PARAMETER["standard_parallel_1",29.5],PARAMETER["standard_p
arallel_2",45.5],PARAMETER["false_easting",0],PARAMETER["false_northing",
0],UNIT["metre",1,AUTHORITY["EPSG","9001"]],AXIS["Easting",EAST],AXIS["Nort
hing",NORTH]]')
```

Figure 4.17 – Land cover raster CRS

Knowing the CRS is only one part that is necessary to georeference the raster file. The second piece of information that is needed is the coordinates of the pixels or grids that make up the raster file. To do this, you can call the .bounds method of the raster object. Calling the bounds method yields the information displayed in *Figure 4.18*.

```
BoundingBox(left=-2361585.0, bottom=259065.0, right=2263815.0, top=3177435.
0)
```

Figure 4.18 – Land cover raster bounds

The bounding box can then easily be converted to a Shapely geometry by passing the output of the bounds method to shapely.geometry.box. That is done by executing the next code block:

```
lc_bbox = shapely.geometry.box(*lc.bounds)
lc_bbox
```

The resulting bounding box is displayed in *Figure 4.19*.

Figure 4.19 – Land cover raster bound as a Shapely geometry

You can now plot the raster file by passing the raster to the show function. Given the size of this raster file, over 15 GB of memory is required to plot the raster file. Given that most personal laptops don't have this amount of memory, you'll need to subset the raster file. To do that, you can create a window using the from_bounds method. You can then pass the window rasterio.open to create a new raster of only the data contained within the window. The following code performs this operation:

```
from rasterio.windows import from_bounds
from rasterio.enums import Resampling

left = -100000.0
right = 100000.0
top = 3177435
bottom = 259065.0

with rasterio.open(data_path + "LCMS_CONUS_v2021-7_Land_Cover_
Annual_2021\\LCMS_CONUS_v2021-7_Land_Cover_2021.tif") as src:
    rst = src.read(1, window=from_bounds(left, bottom, right,
top, src.transform))
    show(rst)
```

The resulting raster plot from this window is displayed in *Figure 4.20*.

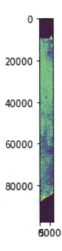

Figure 4.20 – Land cover window plot

Within the show function, you can also pass a color map (cmap) to display the raster using a specific color scale. In the following code, we've changed the bounds and are plotting the raster in greyscale by passing the cmap="Greys" parameter to the show function:

```
left = -150000.0
right = 250000.0
top = 3177435
bottom = 2050000.0

with rasterio.open(data_path + "LCMS_CONUS_v2021-7_Land_Cover_
Annual_2021\\LCMS_CONUS_v2021-7_Land_Cover_2021.tif") as src:
    rst = src.read(1, window=from_bounds(left, bottom, right,
top, src.transform))
    show(rst, cmap="Greys")
```

The resulting output is shown in *Figure 4.21*:

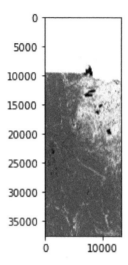

Figure 4.21 – Land cover window in greyscale

To write a raster to a file, you'll again leverage the `rasterio.open()` function. Instead of using the `'r'` mode, this time you'll need to use the `'w'` mode to write out the raster data.

When writing out the raster file, numerous parameters must be predefined and passed to the `.open` function, including the following:

- `driver`: The file format of the resulting raster. GeoTiff is the recommended file format.
- `height`: The number of rows of the raster data.
- `width`: The number of columns of the raster data.
- `bands`: The number of bands. For a single band, the value will be 1. For rasters with red, green, and blue bands, the value will be 3.
- `dtype`: The raster data type represented as a `numpy` data type.
- `crs`: The coordinate reference system, which can be an EPSG code.

Packages enabling spatial analysis and modeling

The prior section focused primarily on packages that enable you to work with and perform operations on spatial data. In this next section, we'll introduce you to packages that allow you to conduct spatial data analysis and modeling.

PySAL

PySAL, or the **Python Spatial Analysis Library**, is a collection of open source packages that support geospatial data science. PySAL's collection of libraries can be broken down into four main categories:

- `Lib`: This is the main library of PySAL, which contains the core backbone architecture for creating spatial indices, working with spatial relationships, and creating what is known as a spatial weights matrix

- `Explore`: Contains libraries that enable you to conduct an exploratory analysis of both spatial and spatiotemporal data

- `Model`: Contains libraries that provide estimations based on spatial relationships present in the data through the use of linear, generalized linear, and non-linear models

- `Viz`: Is a library that enable you to visualize spatial data and the patterns within it to optically detect clusters, hotspots, and outliers

Lib

Inside the Lib component of PySAL exists one package known as `libpysal`. The packages that are contained within the Explore and Model components of PySAL are built upon the infrastructure contained within `libpysal`. `libpysal` contains the following four main modules:

- `libpysal.weights`: This creates a spatial weights matrix

- `libpysal.io`: This handles input and output

- `libpysal.cg`: This handles computational geometry

- `libpysal.examples`: This contains example datasets for use in `libpyal`

Explore

Inside the Explore component of PySAL there exist the following seven packages:

- `esda`: The `esda` package enables you to conduct **exploratory spatial data analysis (ESDA)**; that is traditional **exploratory data analysis (EDA)** with a spatial context. ESDA contains functions for measuring autocorrelation on continuous and binary data at a global and local level.

- `giddy`: **Giddy** stands for **GeospatIal Distribution DYnamics**, and it is an extension of the ESDA package, which enables you to perform operations on spatiotemporal data. Giddy enables measurements for the impact of spatial relationships within temporal data.

- `inequality`: This provides for the analysis of inequities across space and time.

- `momepy`: This is a library for urban morphometrics. **Urban morphometrics** is an emerging field providing an unsupervised and systematic approach for classifying and measuring the urban form.

- `pointpats`: This is a package that enables point pattern analysis.

- `segregation`: This provides for the analysis of segregation across space and time.

- `spaghetti`: This stands for **SPAtial GrapHs: nETworks, Topology, and Inference**. The `spaghetti` package enables the analysis of network-based spatial data and network events.

Model

The Model component of PySAL consists of the following eight packages:

- `access`: This enables the measurement of mismatches in supply and demand across space. Measurement of mismatched supply and demand is critical within a number of industries. As an example, consider the demand on a local hospital network and the location of hospitals serving those patients. In this domain, the appropriate solution to this problem can be the difference between life and death.

- `mgwr`: **MGWR** stands for **multiscale geographically weighted regression**. The `mgwr` package also includes functionality supporting **geographically weighted regression (GWR)**. GWR assumes that the spatial process that is being modeled exists at the same spatial scale, while MGWR relaxes that assumption to model the process at different scales.

- `spglm`: **spglm** stands for **spatial generalized linear model** and is an adaptation of the **generalized linear model (GLM)**. At the time of writing, spglm supports Gaussian GLM, Poisson GLM, quasi-Poisson GLM, and Logistic GLM. The `spglm` package is based upon the `statsmodels` package and is integrated into the `spint` and `mgwr` packages inside PySAL.

- `spint`: **SpInt** stands for **spatial interaction**. The `spint` package provides a collection of techniques for studying processes related to spatial interaction as well as analyzing data related to spatial interaction. At the time of writing, the `spint` package supports unconstrained gravity models, production-constrained models, constrained models, and doubly constrained models.

- Spopt: **spopt** stands for **spatial optimization**, and the `spopt` package supports analysis and models related to facility location optimization and transportation modeling.

- `spreg`: **spreg** stands for **spatial regression**, and the `spreg` package enables simultaneous autoregressive spatial-based regression models.

- `spvcm`: This enables you to estimate **spatially-correlated variance component models** using Gibbs sampling.

- `tobler`: This enables **areal interpolation**, which is the process of making estimates from source polygons onto overlapping but not identical polygons. The `tobler` package also supports **dasymetric mapping**, which is a refinement of choropleth maps using areal interpolation.

Viz

The Viz component of PySAL contains the following three packages:

- `legendgram`: This is a lightweight package for displaying the distribution of the underlying spatial data within the map legend
- `mapclassify`: This is a package for supporting classification based on choropleth maps
- `splot`: This is a lightweight package enabling quick static and dynamic visualizations

At this point, this is all of the detail that we'll go into in regard to PySAL because we will heavily utilize a variety of PySAL packages in *Part 2*, *Exploratory Spatial Data Analysis*, and *Part 3*, *Geospatial Modeling Case Studies*, of this book.

Packages for producing production-quality spatial visualizations

In this section, we'll introduce you to five packages that will enable you to produce spatial visualizations, namely static and interactive maps, which will take your resulting analytical deliverable to the next level. The packages we'll be discussing are `ipyLeaflet`, `folium`, `geoplot`, `geoviews`, and `datashader`.

ipyLeaflet

`ipyLeaflet` is a Python package that enables you to create interactive mapping widgets within the Jupyter Notebook IDE. At its core, `ipyLeaflet` is a connection between the Python IDE and the open sourced, JavaScript-based Leaflet visualization package.

To begin exposing you to the power of `ipyLeaflet`, let's work through creating an interactive map of the attractions in Washington, DC. First, you'll need to import a number of modules from `ipyleaflet` and a handful of other packages, such as GeoPandas, for working with spatial data. To do that, you'll execute the following lines of code:

```
# Importing the packages
from ipyleaflet import (Map, GeoData, basemaps, WidgetControl,
GeoJSON,
  LayersControl, Icon, Marker,basemap_to_tiles, Choropleth,
  MarkerCluster, Heatmap,SearchControl,
  FullScreenControl)
from ipywidgets import Text, HTML
```

```
import geopandas as gpd
import json
```

With the packages imported, you can now begin working with the data. Similar to what was done previously, you'll import the attractions file, geocode the data, and then perform a join to bring back the name of the attraction into the GeoDataFrame. To do this, leverage the following code:

```
import pandas as pd
dc_attractions = pd.read_csv(data_path+ 'DC_Attractions.csv')


from geopandas.tools import import geocode


# Geocode addresses using Nominatim. Remember to provide a
custom "application name" in the user_agent parameter!
Dc_attractions_gpd = geocode(dc_attractions['Address'],
provider='openmapquest', timeout=4, api_key="APIKEY")
dc_attractions_gpd = dc_attractions_gpd.join(dc_
attractions[['Attraction']])


# Adding in lat and lon columns
dc_attractions_gpd['lon'] = dc_attractions_gpd['geometry'].x
dc_attractions_gpd['lat'] = dc_attractions_gpd['geometry'].y


dc_attractions_gpd.head()
```

With the data created, you can now begin creating your `ipyLeaflet` interactive map. First, you'll define the map as `DC_Map` by calling the `Map` function of `ipyLeaflet`. To the `Map` function, you'll pass in a basemap, a center point for the map to focus on, and a zoom level. You can then call the `DC_Map` object to show it in the Jupyter Notebook IDE. The basemap and the center point are important items to pass to the function as they focus the map and the reader on the area of interest with the necessary context as to what they're looking at:

```
DC_Map = Map(
    basemap=basemap_to_tiles(basemaps.Stamen.Toner),
    center=(38.89951498583087, -77.03599825749647),
    zoom=12
)


DC_Map
```

The resulting interactive map currently looks like the screenshot displayed in *Figure 4.22*.

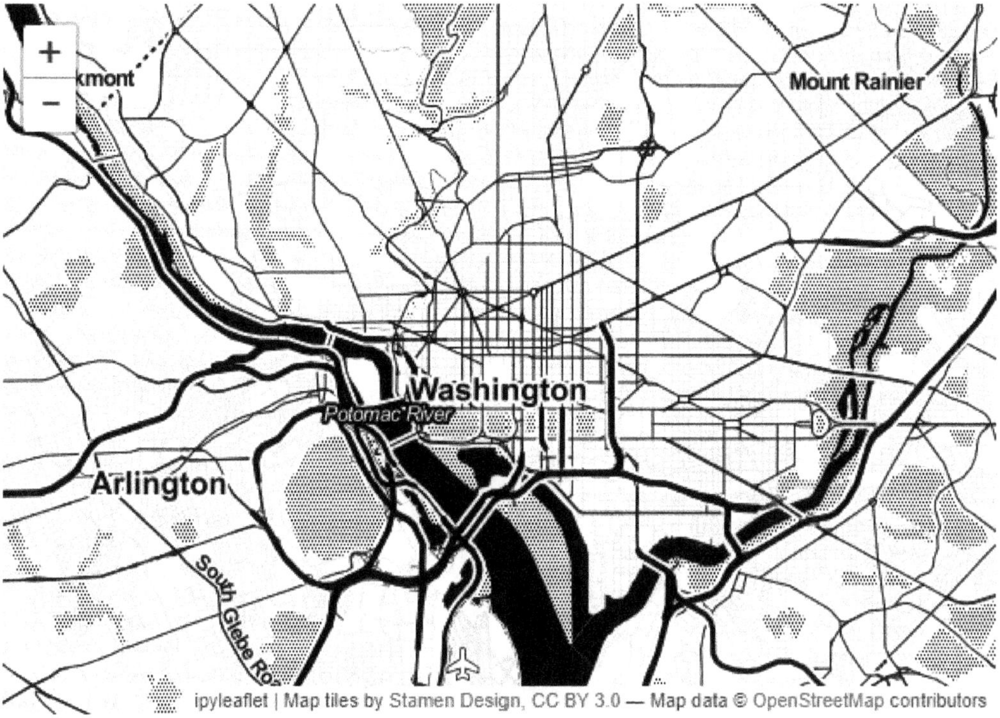

Figure 4.22 – Initial DC map

Leaflet basemaps

There are numerous `Leaflet` basemaps that can be used within your interactive map. Determining the appropriate basemap depends on your analysis. Keep in mind that basemaps exist to provide context and center the end user of the map. For a complete list of basemaps, visit https://ipyleaflet.readthedocs.io/en/latest/map_and_basemaps/ basemaps.html.

To add points that represent the attractions, you'll need to iterate over the rows inside the GeoDataFrame. We'll do this by using the `iterrows` function. As we iterate, you'll pass the latitude and longitude previously defined into the `Marker` function from `ipyLeaflet`. Once the marker is created, you'll use the `add_layer` function to add the marker to the map:

```
# Mapping the attractions

for (index, row) in dc_attractions_gpd.iterrows():
```

```
    marker = Marker(location = [row.loc['lat'], row.
loc['lon']], title=row.loc['Attraction'])
    DC_Map.add_layer(marker)
```

The resulting output is shown in *Figure 4.23*.

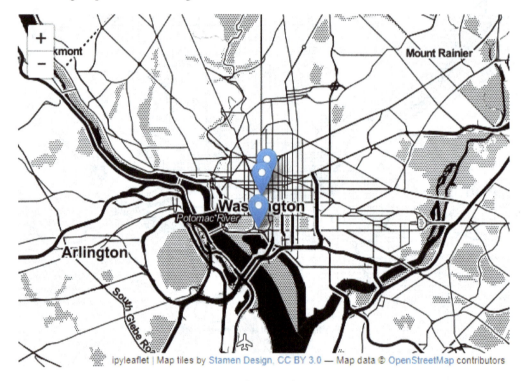

Figure 4.23 – DC map with attractions

You'll notice that as you hover over the plotted points, the name of the attraction shows up. This functionality was enabled previously by passing the `title` parameter to the `Marker` function.

While this is not nearly an exhaustive overview of the `ipyLeaflet` package, it does begin to show you the power within its functions.

Folium

Folium is another visualization package within Python that enables interactive maps. Similar to `ipyLeaflet`, `folium` is built upon the Leaflet visualization package. Folium is better suited for mapping geospatial data that does not change, whereas `ipyLeaflet` is better suited for dealing with user input from the front end that can change the state of the map.

As you may remember, during the section where we discussed GeoPandas, we used the `.plot` method to plot a choropleth map of the world's population. You can now leverage folium to make that map more dynamic and appealing to the end user. To do that, you'll first need to import folium and the required data, as shown in the following code:

```
# importing the folium and geopandas package
import folium

import geopandas as gpd

# Reading in the natural earth lowres data
world = gpd.read_file(gpd.datasets.get_path('naturalearth_
lowres'))
```

With the data ready to go, you can then create a folium map similar to an ipyLeaflet map. To do so, you'll first define a folium map object by calling `folium.Map()`. Next, to produce a choropleth map, you'll call the `folium.Choropleth` function and pass to it the data. By executing the following code, you'll have produced a choropleth map similar to the one shown in *Figure 4.24*:

```
pop_map = folium.Map()



folium.Choropleth(
    geo_data=world,
    name="Population Choropleth map",
    data=world,
    columns=["name","pop_est","gdp_md_est"],
    key_on="feature.properties.name",
    fill_color="YlGn",
    fill_opacity=0.7,
    line_opacity=0.2,
    legend_name="Estimated Population",
).add_to(pop_map)

folium.LayerControl().add_to(pop_map)

pop_map
```

The resulting folium map is displayed in *Figure 4.24*.

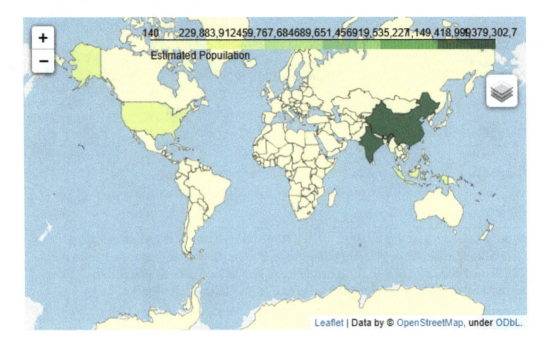

Figure 4.24 – Interactive Folium choropleth map of the world population

Because Folium has largely the same functionality as `ipyLeaflet`, we won't go into too much detail at this time. For more information on the folium package, we recommend that you visit the folium documentation page at: `https://python-visualization.github.io/folium/index.html`.

geoplot

`geoplot` is a static visualization library for geospatial data. The `geoplot` library is an extension of `cartopy` and `matplotlib`. The goal of `geoplot` is to make creating mapping visualizations simple and easy. The ease of use of `geoplot` comes from its high-level mapping API and its native support for projections. For brevity, we'll demonstrate `geoplot` cartogram functions. A **cartogram** is a map in which statistical information is shown in a diagram-based format.

To begin, you'll import `geoplot` while using its alias `gplt`. You'll also need to import the `gcrs` subpackage to work with coordinate reference systems inside `geoplot`. Finally, to work with data we'll again leverage `geopandas`:

```
import geoplot as gplt
import geoplot.crs as gcrs
```

```
import geopandas as gpd
```

```
# Reading in the natural earth lowres data
world = gpd.read_file(gpd.datasets.get_path('naturalearth_
lowres'))
```

With the packages and data imported, you can create your first cartogram. For this cartogram, let's show what the North American continent looks like when the map is scaled to the area of the component country's landmass. To do this, you'll execute the code in the next code block:

```
gplt.cartogram(world[world['continent']=="North America"],
scale=world['area'])
```

The cartogram produced by executing the code is displayed in *Figure 4.25*.

Figure 4.25 – A geoplot cartogram of North American land area

In comparison to the cartogram based on land area, you could instead create a cartogram based on the population estimate of each North American country. To produce this type of visualization, you'll execute the next code block:

```
gplt.cartogram(world[world['continent']=="North America"],
scale='pop_est', projection=gcrs.AlbersEqualArea())
```

The resulting population estimate-based cartogram is displayed in *Figure 4.26*.

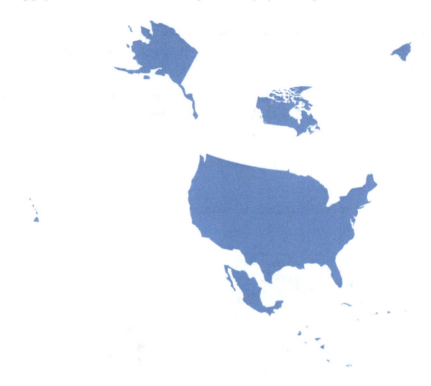

Figure 4.26 – A geoplot cartogram of North America's estimated population

From these two cartograms, you can deduce that there are a lot more people per unit of area measurement in the United States than there are in both Canada and Mexico.

GeoViews

GeoViews is another interactive spatial data visualization library with the goal of making it easy for spatial data scientists to create and explore map-based visualizations. GeoViews is built on top of the HoloViews library, which is beneficial in creating visualizations on highly dimensional data. Spatial visualizations created with GeoViews can be static using matplotlib or dynamic using bokeh to render the data. GeoViews integrates well with GeoPandas, Iris, and Xarray allowing for easy mapping of both vector and raster data structures.

As a simple example, you'll pull from the preloaded features within geoviews to create a map of the Earth. To do this, you'll import geoviews using its standard alias gv, the feature subpackage of geoviews as gf, and cartopy:

```
import geoviews as gv
import geoviews.feature as gf
from cartopy import crs


gv.extension('bokeh', 'matplotlib')
```

With the packages imported, you can then plot an interactive visual of the world using the Bokeh backend by executing the next code block:

```
( gf.ocean * gf.land * gf.coastline * gf.borders).opts(
    'Feature', projection=crs.Geostationary(), global_
extent=True, height=275)
```

The resulting visual, in static form, is shown in *Figure 4.27*.

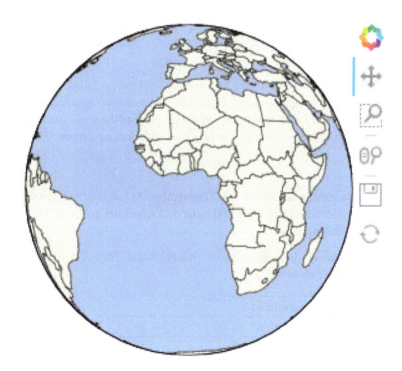

Figure 4.27 – GeoViews map of world countries

You can easily interact with the visualization by using the Bokeh toggles on the right-hand side. These enable you to quickly pan, zoom, and save the visualization to your computer as a `.png` file for use in other mediums, such as a report or presentation.

Datashader

Datashader is relatively new to the world of geospatial data visualizations and helps solve the problem of mapping big geospatial data. As we've mentioned in previous sections, geospatial data is memory intensive to work with. To enable the visualizations to be mapped more efficiently, Datashader breaks the visualization pipeline down into intermediary steps and then uses the Numba and Dask packages to distribute the computation across the CPUs of the machine rendering the visual. By doing this, Datashader enables visualizations of big geospatial data on machines with standard hardware specifications. To learn more about Datashader visit: `https://datashader.org/`.

Reviewing foundational data science packages

In this section, we'll cover a handful of more generalized data science packages, which will be useful within your spatial data science workflows. While these packages are useful for spatial data science, they are not purpose-built for spatial data science workflows, such as the packages covered previously in this chapter.

pandas

pandas is the primary Python package for reading, writing, and manipulating tabular data. As you may recall, the GeoPandas package is built on top of pandas and leverages most of pandas' core functionality. In our coverage of GeoPandas, we did not cover all of the specialized functionality of pandas, which we'll work to cover in this section. Let's start with data structures.

Data structures

The primary pandas data structures are Series and DataFrame. As you may recall from the previous chapter, the GeoPandas' GeoSeries and GeoDataFrame data structures are based on the Series and DataFrame structures within pandas.

A Series data structure is a one-dimensional array that is labeled. The following illustration is an example of a Series with three observations:

Observation1
Observation2
Observation3

Table 4.5 – A pandas Series illustration

Series can store numerous types of data, including the following:

- Integers
- Floating point numbers
- Strings
- Other Python objects

To create a Series, you can define it dynamically or pass one of the following objects:

- A list
- A one-dimensional NumPy array
- A dictionary

The DataFrame data structure is a two-dimensional structure. When you think of DataFrame, you can think of it as a collection of pandas Series. Each column in the DataFrame object holds observations that correspond to a single variable. The data stored in the other columns of a DataFrame may be of the same data type or of a different data type. *Table 4.6* illustrates a DataFrame with multiple data types.

Month	Unit Sales	SalesDollars
1	4	1796.12
2	6	2534.80
3	12	15097.66

Table 4.6 – pandas DataFrame illustration

DataFrame – pandas versus GeoPandas

The primary difference between the GeoPandas DataFrame and the pandas DataFrame is that the GeoDataFrame includes a column that holds the geometry object of the observation. These geometry objects could be a state administrative boundary represented as a polygon, a point that represents a city center, or a line that represents a roadway. There is no concept of geometry within the base pandas DataFrame's data structure.

To create a DataFrame, you can pass one of the following objects:

- A dictionary containing lists, dictionaries, Series, or one-dimensional NumPy arrays
- A Series
- A two-dimensional NumPy array

There are multiple functions within pandas that allow you to read and write tabular data. The functions are listed in *Table 4.7* as follows:

File Type	Operation	
	Reading	**Writing**
CSV	`.read_csv()`	`.to_csv()`
Excel	`.read_excel()`	`.to_excel()`
JSON	`.read_json()`	`.to_json`
Pickle	`.read_pickle()`	`.to_pickle`

Table 4.7 – Reading and writing files with pandas

Now that we've discussed the pandas DataFrame's data structures, let us move on to discussing how to subset a DataFrame in the next section.

Subsetting pandas DataFrames

In order to subset or select a portion of the data in a DataFrame, you'll utilize indexing. Indexing can be done via column, row label, row integer, or slice. See *Table 4.8* for more information.

Procedure	Code
Selecting a column	`df['column']`
Selecting via row label	`df.loc[label]`
Selecting via integer	`df.iloc[integer location]`
Selecting via slice	`Df[1:7]`

Table 4.8 – Subsetting a pandas DataFrame

Another more complex subsetting method comes in the form of Boolean indexing, whereby you use a column's values to subset the DataFrame. We'll walk through an example of Boolean filtering later on in this chapter in the section titled *Combining pandas data structures*.

Methods for exploring pandas DataFrame

There are multiple methods for exploring a pandas DataFrame so that you can begin to understand the data you're working with. You can think of this as the first foray into a deeper process called **exploratory data analysis (EDA)**. EDA is a general process of gathering statistics and visuals describing your dataset so that you begin to understand what it represents and how to use it in your analysis. We'll discuss a lot more about EDA and its sister, **exploratory spatial data analysis (ESDA)** in *Part 2, Exploratory Spatial Data Analysis*. *Table 4.9* details the four most basic data exploration methods in Pandas.

Procedure	Code	Result
View the first n number of rows	`df.head(n)`	Prints the first n number of rows. By default, n is set to 5.
View the last n number of rows	`df.tail(n)`	Prints the last n number of rows. By default, n is set to 5.
View descriptive statistics for the DataFrame	`df.describe()`	Prints descriptive statistics, including the count, mean, minimum, maximum, and quartiles of the data.
View metadata and memory usage of the DataFrame	`df.info()`	Prints the number of columns, the data types, and memory usage. If the verbose parameter is set to `True`, it also prints the number of non-null records.

Table 4.9 – Methods for exploring pandas DataFrames

This concludes our discussion on data exploration within pandas. Next, we'll talk about combining pandas data structures.

Combining pandas data structures

There are three primary functions used to combine pandas Series and DataFrames: `concat`, `merge`, and `join`.

The following list details these functions:

- `pandas.concat()`: Combines data along a prescribed axis. Its primary parameters are the following:

  - `objs`: The Series or DataFrame objects to be combined.

  - `axis`: The axis on which to combine the data. When set to 0, combines data vertically along the index. When set to 1, combined data horizontally.

  - `join`: Used when columns overlap in the objects to be combined. Accepts either `'inner'` or `'outer'`.

- `Pandas.DataFrame.merge()`: Combines data using database-like joins. Its primary parameters are the following:

  - `right`: The right-hand DataFrame.

  - `how`: The type of join to be performed (e.g., `'inner'`, `'outer'`, `'left'`).

  - `on`: The common/key columns to merge upon.

- left_on/right_on: The common key to merge the data upon if the keys are named differently in the left and right DataFrames.

- pandas.DataFrame.join(): Combines data using common columns. Its primary parameters are the following:

  - other: The right-hand DataFrame.

  - on: The common/key columns to merge upon.

  - how: The type of join to be performed. Accepts: 'left', 'right', 'outer', 'inner'.

pandas example

Earlier in this chapter, you conducted an exercise where you geocoded addresses of attractions around Washington, DC. Let's take that data and do some additional workflows with it. Firstly, you may have noticed that when you geocoded the data that the name of the attraction was not in the GeoDataFrame. To correct this, let's use the .merge() function to bring back the name of the attraction. To do that, you'll run the following lines of code:

```
final_gpd = dc_attractions_gpd.merge(dc_
attractions[['Attraction']], how='left',left_index=True, right_
index = True)
final_gpd.head()
```

In this code, you're merging the original pandas DataFrame to the geocoded GeoDataFrame based on the left and right indexes of the input frames. If we had a common identifier in the two frames, we could have used the on parameter instead of the left_index=True and right_index=True parameters. The resulting output of the merged GeoDataFrame is displayed in *Figure 4.28*.

	geometry	address	Attraction
0	POINT (-77.03459 38.90178)	15th Street NW, Golden Triangle, Washington, D...	Washington Monument
1	POINT (-77.03771 38.88772)	Independence Avenue Southwest, Penn Quarter, W...	Smithsonian National Air and Space Museum
2	POINT (-77.03655 38.89772)	1600, Pennsylvania Avenue Northwest, Golden Tr...	White House

Figure 4.28 – Merged GeoDataFrame with the attraction name

Now that you've brought over the attraction name, you can perform a Boolean filtering operation to extract the White House as its own DataFrame. To do that, you'll execute the next line of code:

```
wh = final_gpd[final_gpd['Attraction'] == 'White House']
wh.head()
```

To better understand the Boolean filter, you can execute the following line of code:

```
final_gpd['Attraction'] == 'White House'
```

You'll notice that the results are False, False, True. This is the result of the Boolean filter, which compares every record in the Attraction column to the condition 'White House' to identify which record(s) correspond to the White House.

Finally, let's explore the final_gpd DataFrame to get a better sense of its contents. To do this, we can execute the .info() method on the GeoPandas GeoDataFrame by running the following code:

```
final_gpd.info()
```

The results of this are displayed in *Figure 4.29*.

```
<class 'geopandas.geodataframe.GeoDataFrame'>
Int64Index: 3 entries, 0 to 2
Data columns (total 3 columns):
 #   Column      Non-Null Count  Dtype
---  ------      --------------  -----
 0   geometry    3 non-null      geometry
 1   address     3 non-null      object
 2   Attraction  3 non-null      object
dtypes: geometry(1), object(2)
memory usage: 204.0+ bytes
```

Figure 4.29 – Exploration of the final_gdf GeoDataFrame

In this section, you've learned how to interact with non-spatial data using pandas and its Series and DataFrame structures. You've also learned how to conduct some initial EDA using pandas functions. This section concluded with a discussion about how to combine pandas structures. In the next section, we'll discuss scikit-learn, which is one of the primary modeling packages leveraged in standard data science activities.

scikit-learn

scikit-learn is one of the most robust machine-learning libraries accessible within Python. The scikit-learn library is built upon the SciPy library, which implements a number of algorithms covering methods from optimization to differential equations. scikit-learn expands upon these algorithms, allowing for the development of solutions for classification, regression, clustering, and dimensionality reduction-based problems. In addition to this, it includes functionality for measuring model performance and developing end-to-end modeling pipelines.

As mentioned previously, we will primarily be using PySAL for the geospatial data science modeling use cases covered in *Part 3, Geospatial Modeling Case Studies*. In addition to PySAL, we will leverage aspects of scikit-learn, including its data pre-processing steps, clustering algorithms, and its ability to integrate a **spatial weights matrix**. A spatial weights matrix is a representation of spatial structures via quantification made based on contiguity or distance. We'll elaborate further on this in the next few chapters.

Now that we've covered the geospatial Python packages and a handful of generalized data science packages you'll be leveraging in the remainder of this text, let's transition into a discussion on how to set up your geospatial data science environment.

Summary

In this chapter, we've introduced you to a number of Python packages that you'll frequently leverage in your day-to-day work as a geospatial data scientist. In this chapter, you learned about a variety of packages, such as GeoPandas and Rasterio, for easily reading, manipulating, and writing geospatial vector and raster data structures. In the second part of the chapter, you learned about PySAL and its ecosystem of packages for exploring, modeling, and visualizing geospatial data. The last section of this chapter introduced you to five geospatial packages that enable you to produce production-quality static and interactive geospatial data visualizations.

With these packages, you can now interact with geospatial data and their underlying geometries. Interacting with data is one of the first steps in any data science pipeline. You also now have the skills to produce a variety of geospatial data visualizations, which will be important as you tell the story of your analysis to stakeholders. While we didn't go into too much depth on the spatial analysis packages, we will go into much more detail on this topic, beginning in *Part 2, Exploratory Spatial Data Analysis*.

Part 2:
Exploratory Spatial
Data Analysis

Part 2 of this book focuses on exploratory spatial data analysis. In this section, you'll learn how to craft maps and mapping applications using a variety of Python packages. Through these maps and apps, you'll explore your spatial data and begin your spatial analysis. During this analysis, it is common to theorize hypotheses and begin to test those hypotheses. These hypothesis tests will help you better understand your data and the patterns they present. In the last chapter of this part of the book, you'll be introduced to spatial feature engineering, where you'll derive new attributes from your preexisting data. The lessons learned in this part of the book will enable you to successfully execute a variety of geospatial data science case studies in Part 3.

This part comprises the following chapters:

- *Chapter 5, Exploratory Data Visualization*

- *Chapter 6, Hypothesis Testing and Spatial Randomness*

- *Chapter 7, Spatial Feature Engineering*

5

Exploratory Data Visualization

In *Part 1, The Essentials of Geospatial Data Science*, we provided you with a framework for working with spatial data and progressing through a spatial data science workflow. As a refresher, that framework looked like the one displayed in *Figure 5.1*:

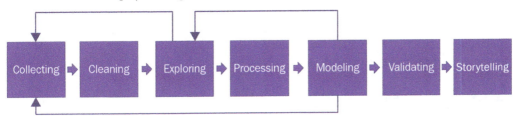

Figure 5.1 – Data science pipeline

In *Part 2, Exploratory Spatial Data Analysis*, the content will focus on the first three steps in the framework: **Collecting**, **Cleaning**, and **Exploring**. For the most part, the collecting step will largely be completed for you, but that will not be the case in the real world. When it comes to the cleaning step, it is often said that data scientists can spend as much as 80-90% of their time cleaning data. Even though we've collected most of the data for you, the data has not yet been cleaned, as learning to clean data—as you can see—is a much-needed and valuable skill to learn.

While the collecting and cleaning stages of the data science pipeline are a focus of this chapter, the primary focus will be on exploring the data, which is *step 3* in the pipeline. In traditional data science, you'll hear this step referred to as **exploratory data analysis** (**EDA**), which is the process of performing initial inquiries of the data in an attempt to discover patterns, test hypotheses, summarize important characteristics of the data, and spot potentially problematic outliers. This process is largely conducted through the production of statistics and the exploration of visualizations.

The geospatial extension of EDA is known as **exploratory spatial data analysis** (**ESDA**). ESDA is a critical component of a spatial data science pipeline that fills in a gap in the more well-known and conventional EDA, which does not focus on the location component or the spatial relationships present in the data. Traditional EDA focuses on understanding the correlation between the variables solely in data space while ESDA relies upon specialized methods to understand the correlation of a variable with respect to its location. We'll dive deeper into this concept throughout *Section 3* of this book.

In this chapter, you'll learn about the following:

- The fundamentals of ESDA

- How ESDA differs from EDA

- How to produce exploratory data visualizations and identify patterns

Technical requirements

For this chapter, you'll leverage the `Chapter 5 - Exploratory Data Analysis` Jupyter notebook, which can be found at this book's GitHub repo using the following link: `https://github.com/PacktPublishing/Applied-Geospatial-Data-Science-with-Python/tree/main/Chapter05`.

The fundamentals of ESDA

Mapmaking, also known as **cartography**, is the first step in ESDA. Mapmaking is a blending of art and science. It is an art form in that you're taking data and representing it in a visual format that is easy to interpret and derive meaning from. Representing data in a visual format is critical for the understanding of both technical and non-technical stakeholders. It is science in that the visuals must be derived from data and they must honor the underlying metadata such as the coordinate reference systems from which they were collected.

Mapmaking is not a standard practice in more traditional EDA, which is traditionally focused on understanding basic statistics of the data such as mean and standard deviation. EDA also focuses on understanding the distribution of the data, dealing with missing data, identifying outliers, and understanding the correlations among variables. In this chapter, you'll work with data and will be conducting the work of traditional EDA while also advancing into the initial stages of ESDA by producing exploratory mapping visualizations. In *Chapter 6*, *Hypothesis Testing and Spatial Randomness*, you'll take your ESDA process one step further by testing theories and better understanding the correlation of your data with respect to its underlying geography.

Before you continue with this chapter, it may be helpful for you to turn back to *Chapter 4*, *Exploring Geospatial Data Science Packages*. In this chapter, you were introduced to a number of different map types, including choropleth maps and cartograms. You were also introduced to a number of different packages from which you can produce map visualizations, including GeoPandas, Folium, and GeoViews.

You'll leverage some of these packages to produce maps that you've already been introduced to as well as learn how to produce new types of maps throughout this chapter.

Example – New York City Airbnb listings

Before we dive deeper into the fundamentals of ESDA, let's take a look at some data. For this exploration, you'll be leveraging the New York City Airbnb data that is collected by Inside Airbnb. Inside Airbnb collects Airbnb data to help support policies and quantify the impact of short-term housing rentals on residential communities.

You can visit Inside Airbnb's website (http://insideairbnb.com/get-the-data) and download the data, which is licensed under a CC 4.0 international license. Once on the website, scroll down or search for New York City and download the listings.csv.gz dataset. Inside Airbnb only stores a rolling 12 months of data, so your results may vary depending upon when you run the analysis.

To begin exploring the data, you'll execute the following steps:

1. First, you need to import the requisite packages including pandas, geopandas, pysal, and splot, to name a few:

    ```
    # Importing the requisite packages
    import pandas as pd
    import geopandas as gpd
    import matplotlib.pyplot as plt
    import pysal
    import splot
    ```

2. Next, you'll import the data and begin to understand the columns in the dataset by running the following code:

    ```
    # Reading in the data
    listings = pd.read_csv(data_path + 'NY Airbnb June 2020\
    listings.csv.gz', compression='gzip')
    print(listings.columns)
    ```

    To better understand the columns, it's helpful to look at the data dictionary for variable definitions. Inside Airbnb's data dictionary can be found here: https://docs.google.com/spreadsheets/d/1iWCNJcSutYqpULSQHlNyGInUvHg2BoUGoNRIGa6Szc4/edit#gid=982310896.

3. Let's now subset the data to keep only a handful of interesting variables, including 'id', 'property_type', 'neighbourhood_cleansed', 'neighbourhood_group_cleansed', 'beds', 'bathrooms', 'price', 'latitude', and 'longitude':

```
# Subsetting the data
listings_sub = listings[['id','property_
type','neighbourhood_cleansed', 'neighbourhood_
group_cleansed','beds','bathrooms',
'price','latitude','longitude']]


listings_sub.head()
```

Running this code produces the table displayed in *Figure 5.2*:

	id	property_type	neighbourhood_cleansed	neighbourhood_group_cleansed	beds	bathrooms	price	latitude	longitude
0	2595	Entire rental unit	Midtown	Manhattan	1.0	NaN	$225.00	40.75356	-73.98559
1	5121	Private room in rental unit	Bedford-Stuyvesant	Brooklyn	1.0	NaN	$60.00	40.68535	-73.95512
2	5136	Entire rental unit	Sunset Park	Brooklyn	2.0	NaN	$275.00	40.66265	-73.99454
3	5178	Private room in rental unit	Midtown	Manhattan	1.0	NaN	$68.00	40.76457	-73.98317
4	5203	Private room in rental unit	Upper West Side	Manhattan	1.0	NaN	$75.00	40.80380	-73.96751

Figure 5.2 – Subset New York City Airbnb data from Inside Airbnb

Now that you've reviewed the data and the data dictionary, you can now turn your focus to conducting EDA and the first stages of ESDA.

Conducting EDA

Let's now explore the data a little bit further in order to better understand basic information such as record counts, data types, and missingness. You can run the `.info()` method on the pandas DataFrame to see how many non-null values there are in each column. Remember that the `.info()` method is one of many pandas methods to explore your data, which were discussed in *Chapter 4, Exploring Geospatial Data Science Packages*. Records with null values indicate an area where the data may need to be cleaned. Output from the `.info()` method is included in *Figure 5.3*:

```
<class 'pandas.core.frame.DataFrame'>
RangeIndex: 37410 entries, 0 to 37409
Data columns (total 9 columns):
 #   Column                        Non-Null Count  Dtype
---  ------                        --------------  -----
 0   id                            37410 non-null  int64
 1   property_type                 37410 non-null  object
 2   neighbourhood_cleansed        37410 non-null  object
 3   neighbourhood_group_cleansed  37410 non-null  object
 4   beds                          36509 non-null  float64
 5   bathrooms                     0 non-null      float64
 6   price                         37410 non-null  object
 7   latitude                      37410 non-null  float64
 8   longitude                     37410 non-null  float64
dtypes: float64(4), int64(1), object(4)
memory usage: 2.6+ MB
```

Figure 5.3 – Subset New York City Airbnb data info

Running this method reveals that there are 37,410 records in the dataset. It also reveals 910 missing records for bedrooms and 37,410 missing records for the bathrooms variable. Given that there is no obvious way to impute the missing values here, we'll go ahead and drop them by using the .drop() method. By specifying the inplace=True parameter, you can drop the records from the original DataFrame without having to create a new DataFrame object:

```
# Cleaning the data
listings_sub.drop(columns=['beds','bathrooms'], inplace=True)
```

You'll also notice that the price variable is stored as an object. In order to perform computations on this column, you'll need to convert it to a float:

```
listings_sub["price"] = listings_sub["price"].replace("[$,]",
"", regex=True).astype(float)
```

To confirm the issues are fixed, you'll display the DataFrame once more, as displayed in *Figure 5.4*:

	id	property_type	neighbourhood_cleansed	neighbourhood_group_cleansed	price	latitude	longitude
0	2595	Entire rental unit	Midtown	Manhattan	225.0	40.753560	-73.985590
1	5121	Private room in rental unit	Bedford-Stuyvesant	Brooklyn	60.0	40.685350	-73.955120
2	5136	Entire rental unit	Sunset Park	Brooklyn	275.0	40.662650	-73.994540
3	5178	Private room in rental unit	Midtown	Manhattan	68.0	40.764570	-73.983170
4	5203	Private room in rental unit	Upper West Side	Manhattan	75.0	40.803800	-73.967510
...	...	...	...	...	...	...	...
37405	640612171111039003	Entire rental unit	Tompkinsville	Staten Island	144.0	40.631656	-74.094878
37406	640629990855220048	Entire townhouse	Williamsburg	Brooklyn	620.0	40.717840	-73.951610
37407	640658348674867448	Private room in rental unit	Gramercy	Manhattan	93.0	40.735540	-73.987880
37408	641072234133563797	Entire rental unit	East Village	Manhattan	462.0	40.729638	-73.987644
37409	641241352138040023	Entire condo	Midtown	Manhattan	113.0	40.752656	-73.972480

Figure 5.4 – Cleaned subset New York City Airbnb data

In this section, you've cleansed the data in the following ways:

- You dropped variables that had a high number of missing values that could not be imputed in a logical way

- You corrected the data type of the price variable, converting it from a string to a float type allowing for future computation

Now that the data is cleaned, let's look at the overall distribution of the `price` variable. First, you'll run the `.describe()` method on the DataFrame, which produces the results shown in *Figure 5.5*:

```
count      37410.000000
mean         190.775221
std          342.491748
min            0.000000
25%           75.000000
50%          125.000000
75%          203.000000
max        12900.000000
Name: price, dtype: float64
```

Figure 5.5 – New York City price statistics

Based on this output, you can see mean nightly price for Airbnbs in New York City is $190.78. The maximum nightly price is $12,900. Based on this information, you can start to understand that the data is skewed. To visually represent this skewness, you can plot a histogram with a kernel density estimation. To do that, you'll run the following code:

```
import seaborn as sns
sns.distplot(listings_sub['price'], kde=True)
plt.show()
```

This produces the visual represented in *Figure 5.6*:

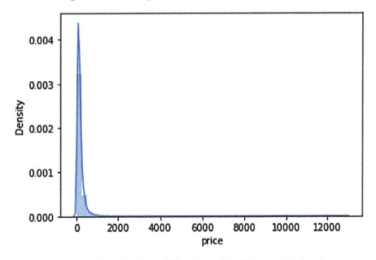

Figure 5.6 – Distribution of New York City Airbnb nightly prices

Now that you've looked at the distribution of the data numerically, let's see how the data distributes over space and begin ESDA.

ESDA

To begin ESDA, you'll first need to convert the pandas DataFrame holding the New York City Airbnb prices into a GeoPandas DataFrame. You'll also want to set its CRS to 4326. Recall from *Chapter 3, Working with Geographic and Projected Coordinate Systems*, that **CRS** stands for **coordinate reference system** and that the **well-known ID (WKID)** 4326-coordinate system is the most standard CRS used in modern computing systems. By checking the data dictionary, you can see that description of the latitude and longitude variables states "*Uses the World Geodetic System (WGS84) projection for latitude and longitude*". This confirms that 4326 is the appropriate CRS for this analysis. The code is illustrated in the following snippet:

```
# Convert the pandas dataframe to a geopandas dataframe
listings_sub_gpd = gpd.GeoDataFrame(listings_sub, geometry=gpd.
points_from_xy(listings_sub.longitude, listings_sub.latitude,
crs=4326))
```

Now that the data is stored as a GeoPandas DataFrame, you can easily plot the observations as a **point map**. A point map uses the latitude and longitude of observations to visualize Airbnb locations on a map. To do this, you'll be using the `geoplot` package we discussed in *Chapter 4, Exploring Geospatial Data Science Packages*. Let's run the following code:

```
import geoplot.crs as gcrs
import geoplot as gplt
ax = gplt.webmap(listings_sub_gpd,projection=gcrs.
WebMercator())
gplt.pointplot(listings_sub_gpd, ax=ax)
```

As a result of the preceding code, you'll produce the map displayed in *Figure 5.7*:

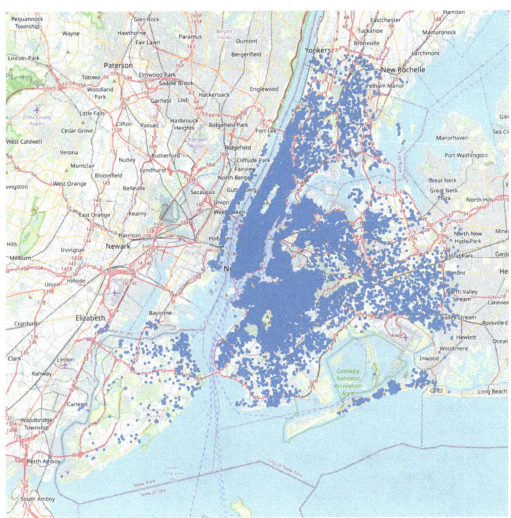

Figure 5.7 – Point map of New York City Airbnb listings

This point map is not highly informative, but it does give you a general sense of the geography from which the data is derived. To start creating better visualizations, let's first convert the point map into a **heatmap**. A heatmap is a spatial visualization that allows you to see the density of points in an area as a raster. To do this, we'll first start by reading in a shapefile associated with New York City's five boroughs:

```
# Borough Boundaries
Boroughs = gpd.read_file(data_path + r"NYC Boroughs\nybb_22a\
```

```
nybb.shp")
Boroughs = Boroughs.to_crs(4326)
Boroughs.plot()
```

The output of running this code is displayed in *Figure 5.8*:

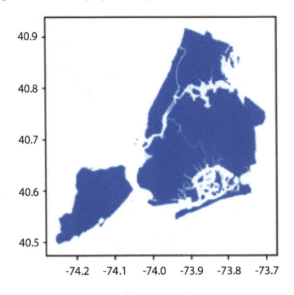

Figure 5.8 – New York City boroughs

Next, you'll create a heatmap using geoplot's kdeplot and polyplot, as detailed in the following code snippet:

```
# Creating a heatmap of Airbnb locations
ax = gplt.kdeplot(
    listings_sub_gpd, shade=True, cmap='Reds',
    clip=Boroughs.geometry,
    projection=gcrs.WebMercator())

# Plotting the heatmap on top of the boroughs for context
gplt.polyplot(Boroughs, ax=ax, zorder=1)
```

The resulting heatmap looks like the one displayed in *Figure 5.9*. From this heatmap, you can see that the highest number of Airbnbs are located in Manhattan and Brooklyn:

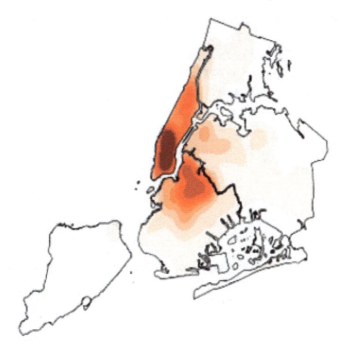

Figure 5.9 – New York City Airbnbs by borough heatmap

In this section, you learned about two new types of maps: a point map and a heatmap, both of which were derived from the raw latitude and longitude of the Airbnb locations. In the next section, you'll aggregate these points up to higher-order geographies.

Converting from point maps to census tract maps

Next, to make the data easier to work with, let's aggregate it up to the census tract level and take the average price of the listings that fall within each census tract. To do that, you'll follow the next few steps:

1. First, you'll need to read in the New York City census tract data from the Census Bureau website and then spatially join and dissolve the listings data:

    ```
    # Reading in the New York Census Tracts
    NY_tracts_path = "https://www2.census.gov/geo/tiger/
    TIGER2021/TRACT/tl_2021_36_tract.zip"
    NY_Tracts = gpd.read_file(NY_tracts_path)
    NY_Tracts = NY_Tracts.to_crs(4326)
    ```

2. Next, you'll subset the census tracts to those that are within the New York **core-based statistical area (CBSA)**:

```
# Subsetting the census tracts to those in the New York
CBSA
cbsa_path = 'https://www2.census.gov/geo/tiger/TIGER2021/
CBSA/tl_2021_us_cbsa.zip'
cbsas = gpd.read_file(cbsa_path)
NY_cbsa = cbsas[cbsas['GEOID']=='35620']


mask = NY_Tracts.intersects(NY_cbsa.loc[620,'geometry'])
NY_Tracts_subset = NY_Tracts.loc[mask]
```

3. Now, it is time to perform the aggregation of the Airbnb locations that fall within the New York City census tracts:

```
# Aggregating the airbnb locations to the NY census
tracts
NY_Tracts_sj = gpd.sjoin(NY_Tracts_subset, listings_sub_
gpd, how='left', op='contains')
NY_Tracts_sj = NY_Tracts_sj[['GEOID','price','geometry']]
NY_Tracts_Agg = NY_Tracts_sj.dissolve(by='GEOID',
aggfunc='mean')
```

4. Finally, you can plot the data using `geoplot` to see how the price is distributed by census tract by running the following code:

```
gplt.choropleth(NY_Tracts_Agg, hue="price",
cmap="Greens", figsize=(60,30), legend=True)
```

The output of this code is displayed in *Figure 5.10*:

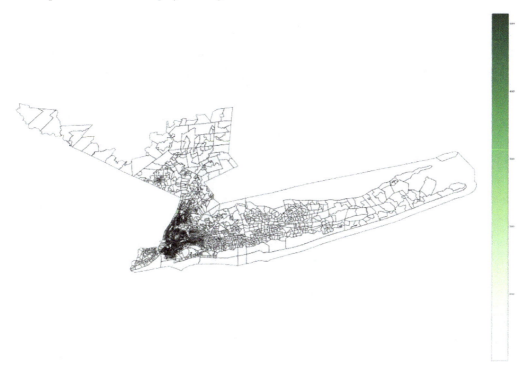

Figure 5.10 – New York City Airbnb nightly price choropleth map

With the map in *Figure 5.10*, we can see that there are a large number of census tracts that do not have any information and therefore are not shaded. This makes it hard to identify the area of interest, which is New York City. Due to there being numerous census tracts without listing information, it is better to produce an interactive map that will allow you, or your end user, to zoom in on areas of interest. You can use the geoviews package with the bokeh extension for this:

```
import geoviews
geoviews.extension("bokeh")

choropleth = geoviews.Polygons(data=NY_Tracts_Agg,
vdims=["price","GEOID"])

choropleth.opts(height=600, width=900, title="NYC Airbnb
Price",
                tools=["hover"], cmap="Greens", colorbar=True,
colorbar_position="bottom")
```

This will produce the interactive map shown in *Figure 5.11*:

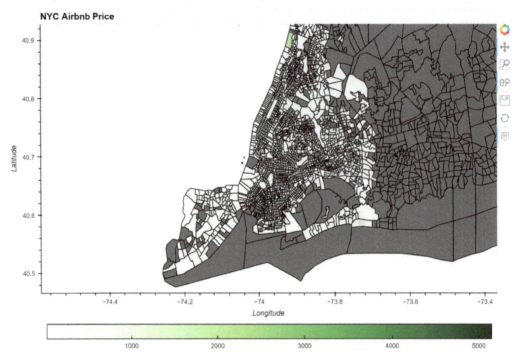

Figure 5.11 – geoviews interactive choropleth map of New York City Airbnb nightly price

Based on the distribution and choropleth visualizations, you now know that there is a long right-hand skew to the average nightly prices for Airbnbs in New York City. To make your visualization a bit more dynamic, let's drop any outliers that are more than one standard deviation away from the mean New York City Airbnb price and then replot the data.

Producing maps without outliers

To produce a map that removes observations greater than one standard deviation from the mean, follow the next few steps:

1. First, let's import the `statistics` package and calculate the mean and standard deviation of the nightly Airbnb price:

```
import statistics

# Calculating the mean and standard deviation
mean_price = statistics.mean(NY_Tracts_Agg['price'].
dropna())
```

```
stdev = statistics.stdev(NY_Tracts_Agg['price'].dropna())


print("The mean Airbnb price is: % s " % (round(mean_
price, 2)))
print("The standard deviation of Airbnb prices is: % s "
% (round(stdev, 2)))
```

2. Next, you'll drop the records that are more than one standard deviation from the mean:

```
# Dropping records that are more than 1 standard
deviation from the mean
NY_Tracts_Agg_filtered = NY_Tracts_Agg[NY_Tracts_
Agg['price'] < mean_price+stdev]
```

3. Finally, plot a more refined choropleth map:

```
choropleth = geoviews.Polygons(data=NY_Tracts_Agg_
filtered, vdims=["price","GEOID"])


choropleth.opts(height=600, width=900, title="NYC Airbnb
Price",
                tools=["hover"], cmap="Greens",
colorbar=True, colorbar_position="bottom")
```

After running the code, you will have a new dataset called NY_Tracts_Agg_filtered where you have removed any observation greater than one standard deviation away from the mean. After removing census tracts with no observations, the dataset has a mean nightly price of $154.64 and a standard deviation of $190.02. The resulting plot can be seen in *Figure 5.12*:

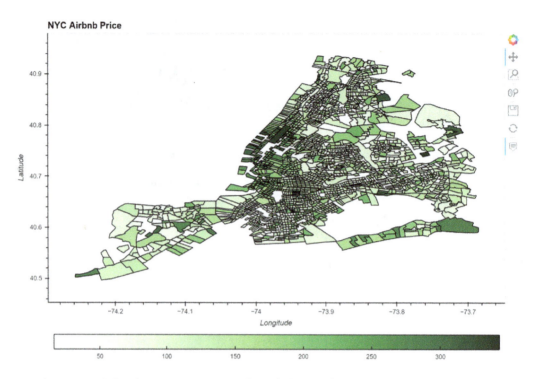

Figure 5.12 – Refined geoviews interactive choropleth map of New York City Airbnb nightly price

By looking at the preceding plot, you can begin seeing some patterns in the data where the nightly price is higher or lower than in other areas. You may also notice that there are clusters of census tracts that all seem to have higher nightly prices compared to those that are further away. These patterns merit further exploration to understand the correlation across space. You'll learn how to do this in *Chapter 6, Hypothesis Testing and Spatial Randomness*.

Summary

In this chapter, you learned about ESDA and how to begin taking it a step further than traditional EDA by creating spatial data visualizations. As an example, you started to work with the New York City Airbnb data by first diving deep into the data to understand some of the issues it has related to missingness, skewness, and ill-formatted data types. After cleaning the dataset up, you learned about three visualizations: point maps, heatmaps, and choropleth maps.

After reviewing the results of the heatmap and the final choropleth map, you began to notice some patterns in the data as it relates to its geographic distribution. You noticed that the highest number of Airbnbs are located in Manhattan and Brooklyn. You also noticed that there was a pattern where certain census tracts were grouped together with higher prices. You'll explore these patterns further in the next chapter.

6

Hypothesis Testing and Spatial Randomness

In *Chapter 5, Exploratory Data Visualization*, you started to understand the first step of **exploratory spatial data analysis** (**ESDA**), which focused on data visualization through the creation of maps derived from the New York City Airbnb dataset. During your work, you noticed that the prices of Airbnb rentals are heavily skewed across New York's geography, with what appeared to be groups of census tracts with higher and lower values in different parts of the city. For reference, take a look at *Figure 6.1*, which represents New York City Airbnb prices as a choropleth map. Areas highlighted by the red circle are groupings of higher values, while areas highlighted by the blue circle are groupings of lower values:

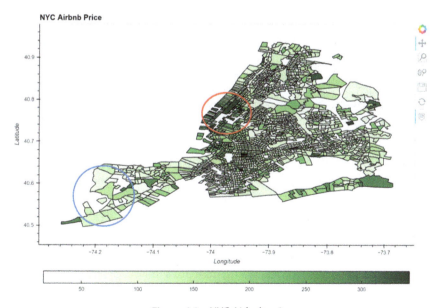

Figure 6.1 – NYC Airbnb prices

This chapter focuses on the second critical part of ESDA, which is testing for spatial structure present within data. Testing for spatial structure is important because if it is present in data, then you'll want to leverage that spatial structure to enhance your downstream analysis. This can be done by using specialized algorithms during a model-building process that can understand patterns from both data and geographic space. By the end of this chapter, you'll have the tools necessary to test whether the New York City Airbnb data exhibits a statistically significant spatial structure.

In this chapter, you'll learn the following:

- How to construct a hypothesis test
- Statistical measures of spatial structures

Technical requirements

For this chapter, you'll leverage the Jupyter notebooks called `Chapter 6 - Spatial Autocorrelation` and `Chapter 6 - Point Pattern Analysis`, which are both stored in the GitHub repo for this book at `https://github.com/PacktPublishing/Applied-Geospatial-Data-Science-with-Python/tree/main/Chapter06`.

Constructing a spatial hypothesis test

In the introduction to this chapter, we mentioned that the second part of ESDA revolves around testing for spatial structure. Before we begin talking about the methods used, let's first discuss what we mean by this term. A **spatial structure** in simplest terms is the presence of a pattern within data across geographic space. Data that has no spatial structure is said to have been generated by an **independent random process** (**IRP**). This IRP result is data that exhibits **complete spatial randomness** (**CSR**). In other literature, you'll often see IRP and CSR used interchangeably. IRP/CSR must satisfy two conditions in order to construct a valid hypothesis test:

- Any observation must have an equal probability of occurring in any location. This is known as a **first-order effect**. As an example, the distribution of an infectious disease will vary across a study area, based on underlying environmental factors.

- The location of an observation is independent and does not impact the location of any other observation. This is known as a **second-order effect**. To continue the previous example on infectious diseases, the intensity of the disease is expected to vary and is dependent on locations where there are infected people, whereas other infections will cluster around the infected people.

> **Understanding first- and second-order effects is important**
>
> A first-order effect means that observations vary from place to place based on changes in the underlying area. A second-order effect means that observations vary from place to place due to interaction effects between observations.

Now that you have an understanding of spatial structures, we can discuss hypothesis testing. **Hypothesis testing** is defined as a statistical test used to determine whether data supports a particular theory or hypothesis. A hypothesis test is broken out into a null hypothesis represented by H_0 and an alternative hypothesis represented by H_a. When testing for CSR, your hypothesis test is as follows:

- H_0: The data is distributed randomly across space

- H_a: The data exhibits a spatial structure and is not randomly distributed

Hypothesis testing relies on specialized methods to understand the correlation of a variable based on its location. In terms of the NYC Airbnb data, one hypothesis test we can perform is called *Moran's I*. Moran's I measures spatial autocorrelation of data based on feature values and feature locations. **Spatial autocorrelation** measures the variation of a variable by taking an observation and seeing how similar or different it is compared to other observations within its neighborhood. Similar to traditional correlation measures, spatial autocorrelation has positive and negative values. Positive spatial autocorrelation results when observations within a neighborhood have similar values, either high-high values or low-low values. Conversely, negative spatial autocorrelation is a result of having low values next to high values, or vice versa.

Figures 6.2 demonstrates, at a high level, what spatial autocorrelation can look like.

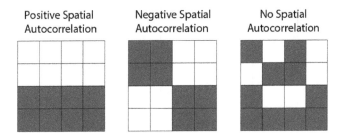

Figure 6.2 – Spatial autocorrelation examples

Spatial autocorrelation is measured in two ways – global and local. **Global spatial autocorrelation** measures the trend in an overall dataset and helps you understand the degree of spatial clustering present. **Local spatial autocorrelation** measures the localized variation in the dataset and helps you detect the presence of hot spots or cold spots. **Hot spots** are localized area clusters with statistically significant high values, and **cold spots** are localized area clusters with statistically significant lower values. Both global spatial autocorrelation and local spatial autocorrelation will be covered in much greater detail later on in the sections titled similarly.

Understanding spatial weights and spatial lags

In order to calculate spatial autocorrelation, you must first start by understanding spatial weights and spatial lags.

Spatial weights

Spatial weights are used to determine the neighborhood for a given observation and are stored in a **spatial weights matrix**. There are three main spatial weights matrices: the rook contiguity matrix, queen contiguity matrix, and **K-Nearest Neighbors (KNN)** matrix. Let's take a look at each of these matrices individually.

The rook contiguity matrix

A **rook contiguity matrix** is created by taking the four nearest neighbors in a north, south, east, and west direction. The name for this method and the choice of nearest neighbors is based on how a rook moves about a chessboard. In order to better understand this, look at *Figure 6.3*, which shows the connected cells for the observation represented by the circle. The rook contiguity matrix would be calculated for this observation based on the cells connected by the dashed lines:

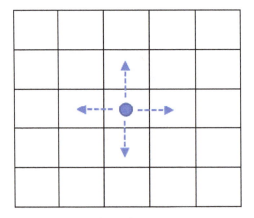

Figure 6.3 – The rook contiguity matrix

The queen contiguity matrix

A **queen contiguity matrix** is created by taking the eight nearest neighbors from every observation, in a similar fashion to how a queen moves about a chessboard. In order to better understand this, look at *Figure 6.4*. The queen contiguity matrix is calculated for the north, south, east, and west cells, represented by the dashed lines, as well as the four corners of the observation, represented by solid lines:

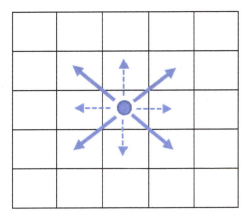

Figure 6.4 – The queen contiguity matrix

The KNN matrix

A KNN matrix is calculated for a given observation based on a set number of nearest neighbors, denoted as k. The number of nearest neighbors to use depends on (and will require a degree of exploration and domain knowledge of) the field or industry that the problem is based upon. *Figure 6.5* represents a KNN matrix using the five closest points.

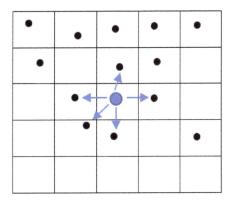

Figure 6.5 – KNN matrix

A quick note on distance calculations

By default, the spatial contiguity matrices are calculated using **Euclidian distance**. Euclidian distance is often referred to as straight-line distance or "as the crow flies" distance. This distance calculation does not take into account natural boundaries, such as rivers and lakes, or man-made obstacles, such as one-way roads or buildings. Depending on the precision needed for your analysis as well as the domain in which you're operating, it may be necessary to base your contiguity matrices on drive distances or walk distances.

Spatial lags

A spatial lag is a byproduct of the spatial weights matrix. A spatial lag is a variable that averages the values of the nearest neighbors, as defined by the spatial weights matrix chosen. For example, if you're working with data pertaining to restaurant sales, you may be interested in restaurant sales in the neighborhood. Therefore, your spatial lag would be the average sales for the restaurants in the neighborhood.

Now that you have an understanding of spatial autocorrelation, spatial weights, and spatial lags, we can take a deeper look at testing for global spatial autocorrelation.

Global spatial autocorrelation

There are a few common statistics used to test for global spatial autocorrelation. The first and most common statistic is Moran's I. Let's take a deep dive into the mathematics of the Moran's I statistic.

Moran's I statistic

The formula for Moran's I is as follows:

$$Moran's\ I\ =\ \frac{N}{S_0}\frac{\sum_{i=1}^{N}\sum_{j=1}^{N}w(i,j)(x_i-\bar{x})\left(x_j-\bar{x}\right)}{\sum_{i=1}^{N}(x_i-\bar{x})^2}$$

where:

- $(x_i-\bar{x})$: The standardized value of the variable of interest
- $\left(x_j-\bar{x}\right)$: The standardized value of the lag of the variable of interest
- $\bar{x}$: The mean of the variable of interest
- $w(i,j)$: The spatial weight between features i and j
- N: The number of features
- S_0: The aggregate of all spatial weights

Global spatial autocorrelation – the NYC Airbnb example

For this example, you'll continue to use the NYC Airbnb dataset discussed in the last chapter. You'll follow the following steps to complete the exercise and test for global spatial autocorrelation:

1. Start by importing the requisite packages for this exploration:

    ```
    # Importing the requisite packages
    import pandas as pd
    ```

```
import geopandas as gpd
import matplotlib.pyplot as plt
import pysal
import splot
import matplotlib.pyplot as plt
import seaborn
from pysal.viz import splot
from splot.esda import plot_moran
import contextily
from IPython.display import display, Markdown, display_
latex, display_markdown, display_html
```

2. With the packages imported, you'll now need to prepare the data similar to how it was treated in *Chapter 5, Exploratory Data Visualization*. The code to do this is included in the notebook for this chapter.

3. With data cleaning already completed, the focus now turns to developing the spatial hypothesis test. To do this, you'll import a few additional packages:

```
from pysal.explore import esda
from pysal.lib import weights
from numpy.random import seed
```

4. With the `weights` module from `pysal`, you'll construct a spatial weights matrix of the listings data. For this example, we're going to use the queen-based spatial weights matrix. We're also going to **row-standardize** the weights matrix. Row standardization occurs by dividing the weight for a feature by the sum of all neighbor weights for that same feature. It is generally recommended that this process be applied any time there is potential bias due to the sampling construct or the aggregation process:

```
# Generate W from the GeoDataFrame
# Need to convert this to polygon data
w = weights.Queen.from_dataframe(NY_Tracts_Agg_filtered)

# Row-standardization
w.transform = 'R'
```

5. Next, you'll calculate the spatial lag. You'll also need to standardize the variable of interest and its spatial lag. Standardizing the values helps you see values above or below the mean easier when plotting the Moran's I scatter plot:

```
# Calculate the spatial lag
NY_Tracts_Agg_filtered['price_lag'] = weights.spatial_
lag.lag_spatial(
    w, NY_Tracts_Agg_filtered['price']
)


# Standardize the variable of interest and the lag
variable
NY_Tracts_Agg_filtered['price_std'] = (NY_Tracts_Agg_
filtered['price'] - NY_Tracts_Agg_filtered['price'].
mean())
NY_Tracts_Agg_filtered['price_lag_std'] = (NY_Tracts_Agg_
filtered['price_lag'] - NY_Tracts_Agg_filtered['price_
lag'].mean())
```

6. You can then plot the spatial weights and spatial lag against one another to produce the Moran's I scatter plot:

```
f, ax = plt.subplots(1, figsize=(10, 10))
seaborn.regplot(
    x='price_std', # variable of interest
    y='price_lag_std', # spatial lag
    ci=None, # suppress the plotting of the confidence
interval
    data=NY_Tracts_Agg_filtered, # dataset
    line_kws={'color':'r'}
)

ax.axvline(0, c='k', alpha=0.8)
ax.axhline(0, c='k', alpha=0.8)

ax.set_title('Moran Plot - NYC Airbnb Price')
ax.set_xlabel('Standardized Price')
ax.set_ylabel('Standardized Price Lag')
plt.show()
```

The resulting scatter plot is displayed in *Figure 6.6*.

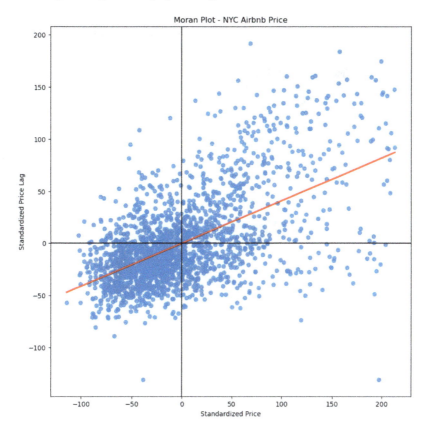

Figure 6.6 – Moran's I plot of NYC Airbnb prices

The points in the upper-right quadrant have a positive global spatial autocorrelation, while the points in the lower-left quadrant have a negative spatial autocorrelation. A line of best fit, shaded red, is overlayed on the plot to show the strength of the relationship. Here, the relationship appears to be reasonably strong, as the points are distributed above and below the line of best fit to roughly the same degree.

7. You can now calculate the Moran's I statistic by using the `Moran` module, as shown in the next code section:

```
morans_stat = esda.moran.Moran(NY_Tracts_Agg_
filtered['price'], w)
display(Markdown(f""""**Morans I:** {morans_stat.I}"""))
display(Markdown(f""""**p-value:** {morans_stat.p_
sim}"""))
```

The resulting Moran's I statistics is ~0.41, with a *p*-value of 0.001. This indicates that there is a statistically significant spatial relationship present in the NYC Airbnb price data.

The `Moran` module also makes it easy to plot the Moran's I scatter plot by calling the `plot_moran()` function:

```
plot_moran(morans_stat)
```

The resulting plot is displayed in *Figure 6.7*.

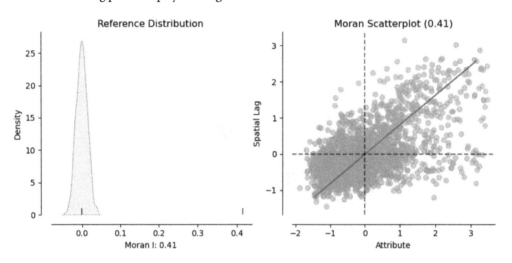

Figure 6.7 – Moran's I plot of NYC Airbnb prices using plot_moran

8. There are other variables of interest in the NYC Airbnb dataset that may exhibit a spatial structure. In the next step, you'll explore the spatial structure of the `accommodates` and `bedrooms` variables. To do that, you'll execute the following piece of code:

```
import numpy as np
from esda.moran import Moran

# Set seed for reproducibility
np.random.seed(54321)

# Set the variables of interest
variables_of_interest = ['accommodates','bedrooms']

for voi in variables_of_interest:
    morans_stat = esda.moran.Moran(NY_Tracts_Agg_
```

```
filtered[voi], w)
    display(Markdown(f""""**Morans I for {voi}:** {morans_
stat.I}"""))
    display(Markdown(f""""**p-value for {voi}:** {morans_
stat.p_sim}"""))
    plot_moran(morans_
stat)     display(Markdown(f""""**Morans I for {voi}:**
{morans_stat.I}"""))
    display(Markdown(f""""**p-value for {voi}:** {morans_
stat.p_sim}"""))
    plot_moran(morans_stat)
```

For the `accommodates` variable, the Moran's I statistic is 0.11 with a p-value of 0.001. *Figure 6.8* shows the Moran scatter plot for the `accommodates` variable.

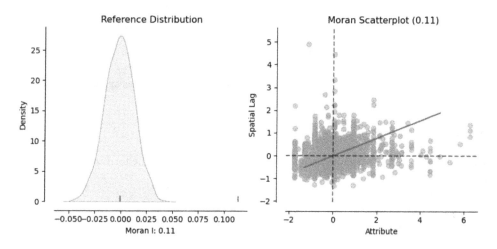

Figure 6.8 – Moran's I plot of the NYC Airbnb accommodates variable

For the `bedrooms` variable, the Moran's I statistic is 0.10 with a *p*-value of 0.001. *Figure 6.9* shows the Moran scatterplot for the `bedrooms` variable.

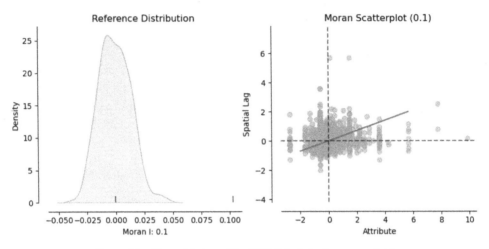

Figure 6.9 – Moran's I plot of the NYC Airbnb bedrooms variable

In addition to the Moran's I statistic, there are other common measures of global spatial autocorrelation. Geary's C is one such measure.

Geary's C statistic

Let's take a deep dive into the mathematics behind Geary's C statistic. The formula for Geary's C is as follows:

$$Geary's\ C = \frac{(N-1)\sum_{i=1}^{N}\sum_{j=1}^{N}w(i,j)(x_i-x_j)^2}{2S_0 \quad \sum_{i=1}^{N}(x_i-\bar{x})^2}$$

where:

- $(x_i - \bar{x})$: The standardized value of the variable of interest
- $(x_j - \bar{x})$: The standardized value of the lag of the variable of interest
- $\bar{x}$: The mean of the variable of interest
- $w(i,j)$: The spatial weight between features i and j
- N: The number of features
- S_0: The aggregate of all spatial weights

Now that we've reviewed the mathematics behind the Geary's C statistic, let's jump back to our example with the NYC Airbnb data and calculate Geary's C for our three variables.

To calculate the Geary's C statistic for the three variables, the following code can be executed:

```
# Set the variables of interest
variables_of_interest = 'price','accommodates','bedrooms']

for voi in variables_of_interest:
    geary_c = esda.geary.Geary(NY_Tracts_Agg_filtered[voi], w)
    display(Markdown("""**Geary's C for {voi}:**
{geary_c.C"""))
    display(Markdown("""**p-value for {voi}:** {geary_c.p_
sim"""))
```

The values for Geary's C are represented in *Table 6.1*.

Variable	Geary's C	p-value
price	0.58	0.001
Accommodates	0.87	0.001
Bedrooms	0.89	0.001

Table 6.1 – The Geary's C statistics

For each of the statistics calculated, all three of the variables exhibited a spatial structure that is statistically significant, as shown in the summary represented in *Table 6.1*.

It is worth noting that each of these statistics will vary based on the spatial weights matrix that you choose to construct. In these examples, we've used the queen contiguity matrix. Had we chosen to use the rook or KNN-based matrices, the statistic would likely have changed.

This section has largely focused on global spatial autocorrelation. In the next section, we'll dive deeper into local spatial autocorrelation.

Local spatial autocorrelation

In the previous section, we focused on global spatial autocorrelation. In this section, we will introduce you to local spatial autocorrelation. Local spatial autocorrelation measures the relationship between each observation and its localized surroundings. Compared to global spatial autocorrelation, local spatial autocorrelation does not return a single value but instead returns values per observation.

To begin, let's discuss **Local Indicators of Spatial Associations (LISAs)**. LISAs are spatial statistics that are derived from global spatial statistics and calculate local cluster patterns, also known as spatial outliers. These spatial outliers are unlikely to appear if the assumption of spatial randomness was true.

LISAs must conform to two requirements:

- LISAs for each and every observation represent the degree of spatial clustering of similar values in that observation's neighborhood
- The sum of LISAs across an observation must be proportional to the associated global indicator from which the LISAs are derived

One well-known LISA is Local Moran's I, which describes localized spatial clustering in terms of high and low values. Local Moran's I can be calculated using the `Moran_Local()` function.

Local spatial autocorrelation – the NYC Airbnb example

Let's jump back into our example with the NYC Airbnb data to test for local spatial autocorrelation:

1. Leverage the `Moran_Local()` function to calculate the LISA for the `price` variable:

```
price_lisa = esda.moran.Moran_Local(NY_Tracts_Agg_
filtered''price''], w)
```

2. To better understand the output of the Local Moran's I statistic, you can plot the distribution of the `Is` attribute of the `price_list` object, which is shown in *Figure 6.10*:

```
# Draw KDE line
f, ax = plt.subplots(1, figsize=(10, 10))
seaborn.kdeplot(price_lisa.Is, ax=ax)

plt.show()
```

The resulting plot is as follows:

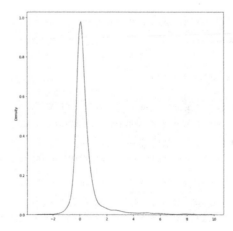

Figure 6.10 – A Moran's Local I distribution of NYC Airbnb prices

From the distribution, you can see that there is a massive spike in the data around **0** with a long right tail. This is mainly due to the presence of a large number of observations with positive spatial autocorrelation, which is in line with what we discovered from the global measures.

Let's now calculate some additional visualizations to help us better understand the output of the Local Moran's I statistic:

1. First, let's create a choropleth map of the Local Moran's I values:

```
f, ax = plt.subplots(1, figsize=(10, 10))


# Create a new column with the values from the Moran's
Local LISAs
NY_Tracts_Agg_filtered.assign(
    ML_Is=price_lisa.Is

# Plot choropleth of local statistics
).plot(
    column='ML_Is',
    cmap='vlag',
    scheme='quantiles',
    k=4,
    edgecolor='white',
    linewidth=0.1,
    alpha=0.75,
    legend=True,
    ax=ax
)

plt.show()
```

This map is displayed in *Figure 6.11*:

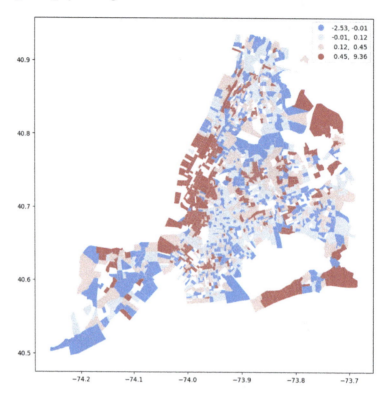

Figure 6.11 – A choropleth map of the NYC Airbnb price Local Moran's Is

2. Next, you can plot a map of all the observations, assigning each observation to its quadrant: **High-High (HH)**, **High-Low (HL)**, **Low-High (LH)**, and **Low-Low (LL)**. For this first example, we'll force the function to give us quadrant values for all observations, regardless of whether those observations are statistically significant or not:

```
# import additional package esda from splot:
from splot import esda as esdaplot


# Plot a map assigning each observation with its quadrant
value HH, HL, LH, LL
f, ax = plt.subplots(1, figsize=(10, 10))
esdaplot.lisa_cluster(price_lisa, NY_Tracts_Agg_filtered,
p=1, ax=ax);


plt.show()
```

Figure 6.12 is the map produced by the previous code cell, which shows the concentrations of **HH**, **HL**, **LH**, and **LL** clusters based on the nightly Airbnb price.

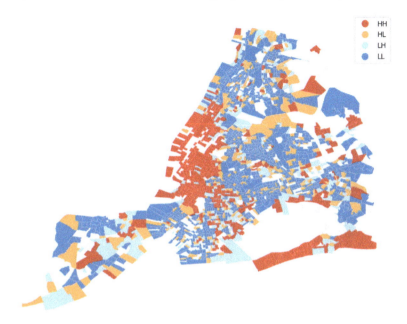

Figure 6.12 – The NYC Airbnb price Local Moran's I quadrant map

3. Next, let's look at which of these points have statistically significant values at an alpha of 0.05. We'll assign this significance level to the object called `alpha`:

```
f, ax = plt.subplots(1, figsize=(10, 10))


# First, we need to find out which observations are
significant
alpha = 0.05
labels = pd.Series(
    1 * (price_lisa.p_sim < alpha), # 1: Indicates
significance at alpha of .05 and 0 indicates
insignificant values
    index=NY_Tracts_Agg_filtered.index

# Recoding 1 to be "Significant and 0 to be
"Insignificant"
).map({1: 'Significant', 0: 'Insignificant'})
```

```
# Creating a new column with the labels for significance
called ML_Sig
NY_Tracts_Agg_filtered.assign(
    ML_Sig=labels

# Plotting a map of the insignificant values
).plot(
    column='ML_Sig',
    categorical=True,
    k=2,
    cmap='vlag',
    linewidth=0.1,
    edgecolor='white',
    legend=True,
    ax=ax
)
```

The result of running the code in the previous code section is the choropleth map of significant and insignificant values, as shown in *Figure 6.13*.

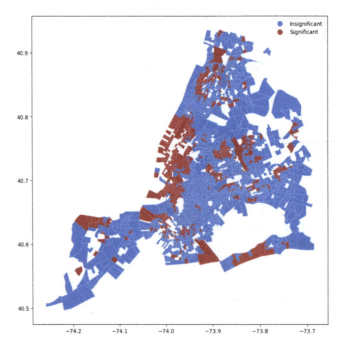

Figure 6.13 – A choropleth map of NYC Airbnb price Local Moran's I significance

4. Finally, we'll plot a choropleth map showing the quadrant values for the significant observations and a fifth value representing the insignificant observations:

```
# Plot one final map using the alpha of 0.05
f, ax = plt.subplots(1, figsize=(10, 10))

esdaplot.lisa_cluster(price_lisa, NY_Tracts_Agg_filtered,
p=alpha, ax=ax);

plt.show()
```

Figure 6.14 shows the choropleth map with quadrant labels for significant observations.

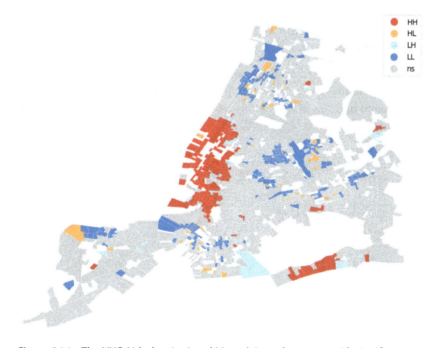

Figure 6.14 – The NYC Airbnb price Local Moran's I quadrant map with significance

The prior two sections have thoroughly covered both global and local spatial autocorrelation. In the next section, we'll introduce you to point pattern analysis, which is another way to examine the spatial structure in your data.

Point pattern analysis

Up until now, this chapter has solely focused on spatial autocorrelation. Spatial autocorrelation is just one spatial structure that can be tested. Another spatial hypothesis test falls within the domain of **point pattern analysis**. Point pattern analysis centers around the patterns present within point data instead of the attributes associated with the point data.

Studying the patterns present in point data is very common in the study of infectious diseases. As we discussed at the start of this section with respect to first- and second-order spatial effects, diseases are often clustered together around infected individuals or other infectious origins. One of the earliest uses of maps to identify the origin of an infectious disease was Dr. John Snow's famous cholera map. While Dr. Snow didn't have the statistics or technology that we have today, he was able to use maps and spatial data to identify that the infection originated from contaminated drinking wells in London. If you're unfamiliar with Dr. Snow's work, you can visit this resource from the Royal College of Surgeons of England: `https://www.rcseng.ac.uk/library-and-publications/library/blog/mapping-disease-john-snow-and-cholera/`.

Today, we have more sophisticated measures and technology to understand the patterns that are present in our data. With these measures, you will often focus on the degree of aggregation in the point data, as measured in terms of dispersion and clustering. *Figure 6.15* shows point data that is randomly distributed.

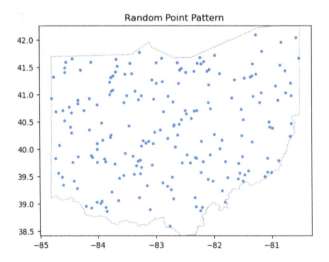

Figure 6.15 – Randomly distributed point data

Figure 6.16 shows point data with a large degree of clustering.

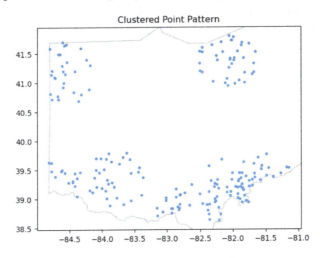

Figure 6.16 – Clustered point data

At an initial glance, *Figure 6.15* may appear to be somewhat clustered and not look like randomly derived data. This is because the human brain is built to try and detect patterns, and it wants to see a pattern in this data. However, this data is generated by taking random points from a Poisson distribution. This is one of the reasons we conduct point pattern analysis and why testing the H_0 value of complete spatial randomness is so important.

Ripley's alphabet functions

There are many statistics for measuring the pattern or lack thereof in point data. Most of these statistics are within Ripley's alphabet of functions. In this section, we'll cover Ripley's G and Ripley's K. There are many other statistics that are within Ripley's alphabet functions that are not covered in this text. For more information on them, we encourage you to read *Statistical Inference for Spatial Processes* by Brian Ripley.

To best understand how to apply Ripley's alphabet functions, it is best to start with an example. For this example, we're going to utilize a dataset of Dollar General retail store locations within Ohio. This data was sourced using the OpenStreetMap API and is available under the Open Data License. To learn more, visit `https://www.openstreetmap.org/copyright`:

1. To begin this exercise, you'll first need to import the requisite packages:

    ```
    ### Importing the requisite packages

    #Data management
    ```

```
import geopandas as gpd
import numpy as np
import pandas as pd

# Visualization
import seaborn
import contextily
import matplotlib.pyplot as plt

# Spatial Statistics
from pointpats import distance_statistics, QStatistic,
random, PointPattern
```

2. Next, you'll need to read the CSV file storing the store locations into a pandas DataFrame and convert it to a GeoPandas GeoDataFrame:

```
### Setting the file path
data_path = r'YOUR FILE PATH'

# Reading in the data from the path
locs_pdf = pd.read_csv(data_path + 'OSM_
DollarGeneralLocs.csv')

# Converting the pandas dataframe into a geopandas
geodataframe
locs_gdf = gpd.GeoDataFrame(
    locs_pdf, geometry=gpd.points_from_xy(locs_pdf.X,
locs_pdf.Y),
    crs="EPSG:4326"
```

With the data imported, you can call the `GeoPandas.plot()` function to create an initial scatter plot of the data. The resulting visualization is displayed in *Figure 6.17*:

```
# Create an initial visualization of the data
f, ax = plt.subplots(1, figsize=(8, 8))
ohio_st.plot(ax = ax,color=None, zorder=1)
locs_gdf.plot(ax = ax, zorder=2, color='black',
markersize=8)
```

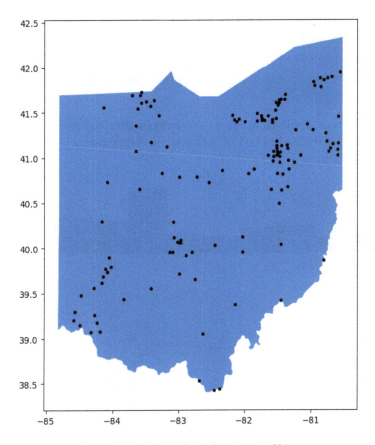

Figure 6.17 – A plot of store locations in Ohio

This plot doesn't provide too much context for your analysis.

3. To provide a little more context, call Seaborn's `jointplot()` and pass to it a contextily `basemap` to provide as a reference layer:

```
# Create a better visualization
joint_axes = seaborn.jointplot(
    x=locs_gdf.geometry.x,
    y=locs_gdf.geometry.y,
    data=locs_gdf,
    s=5,
    height=7,
color='k');
```

```
contextily.add_basemap(
    joint_axes.ax_joint,
    crs="EPSG:4326",
    source=contextily.providers.Stamen.TonerLite
);
```

The resulting visual, displayed in *Figure 6.18*, shows the distribution of store locations across Ohio. At the top and right-hand side of the map is the distribution of points at various latitude and longitude degrees:

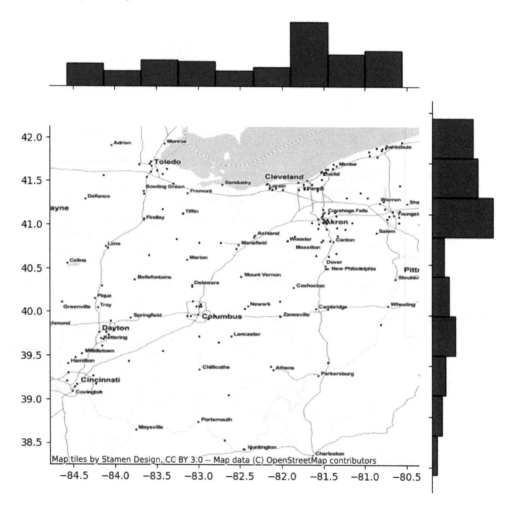

Figure 6.18 – Store locations in Ohio displayed via jointplot()

Through studying the map, you begin to see that there are many stores located around cities such as Dayton, Columbus, Cleveland, and Akron. To see whether these stores are clustered together in a spatially significant way, you can begin leveraging some of Ripley's statistics.

We'll start with **Ripley's G**, which is a cumulative function describing the distribution of distances between nearest neighbor points. By comparing the cumulative function from Ripley's G with a point dataset that is generated from a spatially random process, you can determine whether the point data you're studying expresses a spatial pattern. In addition to Ripley's G, **Ripley's F** and **Ripley's J** also leveraged cumulative distribution functions:

1. Coding Ripley's G in Python is relatively simple. All you need to do is call the g_test() function from pointpats, which you imported earlier:

```
g_test = distance_statistics.g_test(
    locs_gdf[['X','Y']].values, support=40, keep_
simulations=True
)
```

2. You can then plot Ripley's G on top of a simulated CSR dataset by running the following code:

```
plt.plot(g_test.support, np.median(g_test.simulations,
axis=0),
        color='k', label='Simulated Data')
plt.plot(g_test.support, g_test.statistic,
        marker='x', color='orangered', label='Observed
Data')
plt.legend()
plt.xlabel('Distance')
plt.ylabel('Ripleys G Function')
plt.title('Ripleys G Function Plot')
plt.show()
```

Figure 6.19 shows the resulting Ripley's G plot. The orange line represents the cumulative distance function from the store locations dataset. The black line represents a simulated CSR distribution. Between distances of 0 and ~.25, the observed data rises much more rapidly than the simulated data from a CSR process. From this, we can deduce that the store location data has a significant spatial pattern.

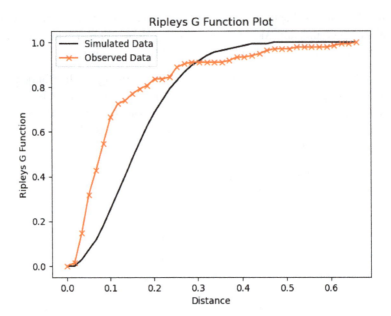

Figure 6.19 – The Ripley's G plot

Next, we'll test the store locations dataset with the **Ripley's K** function. In contrast to the Ripley's G function, Ripley's K considers all distances and not just those of the closest neighbor. **Ripley's L** is another such function that considers all distances.

3. Call the k_test() function from pointpats to run Ripley's K:

```
k_test = distance_statistics.k_test(locs_gdf[['X','Y']].
values, keep_simulations=True)
```

4. You can then plot the output of Ripley's K against the simulated data from a CSR process, similar to what you did when testing Ripley's G:

```
plt.plot(k_test.support, k_test.simulations.T, color='k',
alpha=.01)
plt.plot(k_test.support, k_test.statistic,
color='orange')

plt.scatter(k_test.support, k_test.statistic,
           cmap='viridis', c=k_test.pvalue < .05,
           zorder=4
           )
```

```
plt.xlabel('Distance')
plt.ylabel('Ripleys K Function')
plt.title('Ripleys K Function Plot')
plt.show()
```

The resulting plot is displayed in *Figure 6.20*.

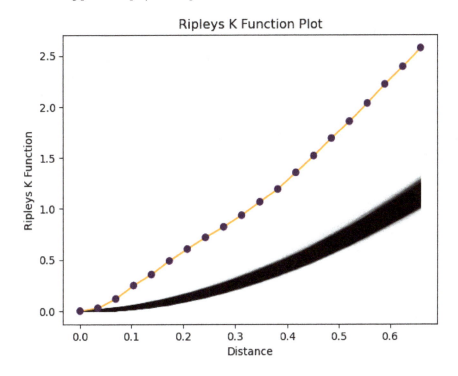

Figure 6.20 – The Ripley's K plot

Here again, we can see that the observed data is well above that of the simulated data, which confirms again that this data is from a process that is not spatially random. It makes sense that the store locations exhibit a pattern and are not the result of a spatially random process. This is because the executives leading this business are likely looking for locations with a given set of geodemographic characteristics known to be indicative of positive business performance. It would be unwise of them, and likely unprofitable, to place store locations randomly around Ohio.

Summary

In this chapter, you learned how to construct multiple hypothesis tests. Each hypothesis test was conducted with the null hypothesis (H_0) as CSR. The alternative hypothesis (H_a) in each case was that data exhibited a non-random, spatially significant relationship.

For data where the attribute features were important, you conducted tests of global spatial autocorrelation, leveraging Moran's I and Geary's C. You also learned how to identify known spatial outliers and hot and cold spots through the use of LISAs.

At the end of the chapter, you learned how to conduct hypothesis testing on point data where the distribution of the points themselves was of interest. The tests you conducted here were based on Ripley's alphabet functions, which looked at the nearest neighbor distance distributions and full-distance distributions.

Now that you've thoroughly explored this data, it can be leveraged in future exercises. We hope you're excited to begin engineering some features in *Chapter 7, Spatial Feature Engineering*, which is up next.

7
Spatial Feature Engineering

As we kick off this chapter, it's helpful for us to recall the data science pipeline and identify where we are within it at this stage of the book. Take a look at *Figure 7.1*, the data science pipeline.

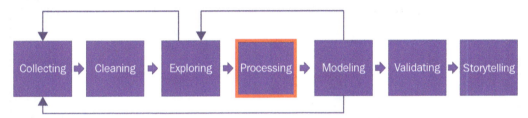

Figure 7.1 – Data science pipeline

In *Chapter 5, Exploratory Data Visualization*, and *Chapter 6, Hypothesis Testing and Spatial Randomness*, you focused on exploring some datasets and testing for spatial relationships. Recall that in the New York Airbnb dataset, you identified that there was spatial autocorrelation present at both a global and local level. In this chapter, you'll be focused on the part of the data science pipeline that we call **processing**, as highlighted in red in *Figure 7.1*. Other texts may refer to this step in the data science pipeline as **data engineering** or **feature engineering**. Within this step of the pipeline, your focus is on manipulating and transforming raw data into features that are best suited for your modeling exercise.

While this is the first time in this book that we're dedicating ample space to discuss this topic, it is not the first time you've seen this step in action. In fact, you've already done a lot of feature selection and have also done some initial feature engineering. Recall the following:

- In *Chapter 4, Exploring Geospatial Data Science Packages*, you created geographic features from addresses representing important tourist sites in Washington, DC. We call this **spatial feature engineering**.

- In *Chapters 5* and *6*, you selected variables of interest in the Airbnb analysis. Those variables included `price`, `beds`, `bedrooms`, and `accommodates`, to name a few. This represents variable selection in action.

In this chapter, we will dive much deeper into the processing step and will be contributing significant focus on the topic of spatial feature engineering and the methods by which this is done.

By the end of this chapter, you will have an understanding of the following:

- What spatial feature engineering is
- How to derive new spatial features from existing spatial data

Technical requirements

For this chapter, you'll leverage the Jupyter notebook called `Chapter 7 - Spatial Feature Engineering`, which is stored in the GitHub repository for this book at `https://github.com/PacktPublishing/Applied-Geospatial-Data-Science-with-Python/tree/main/Chapter07`.

Defining spatial feature engineering

As we mentioned at the beginning of this chapter, feature engineering refers to the manipulation and transformation of raw data into features that are best suited to your analytical exercise. In data science, feature engineering can take many forms, including the following:

- Filling missing values, leveraging expert intuition, or various machine learning-based approaches
- Scaling and normalization, whereby the range and center of data are adjusted to help train models and allow easier interpretation later on
- Feature encoding, whereby categorical data is converted to binary `True` or `False` representation across multiple columns

Spatial feature engineering is very similar to the approaches taken in more general data science. It is the process of creating, or engineering, new and additional information from raw data using geographic context and knowledge. Engineering new features can be done by connecting data from two or more datasets using geography or from within existing spatial data, where the underlying spatial structure is used to create additional features.

Spatially engineered features come in two main forms: **summary spatial features** and **proximity spatial features**. Let's define both of these forms and go through a few examples:

- Summary spatial features are created based on a preexisting spatial relationship between an observation and another feature. Examples include the following:

- Descriptive statistics derived from features within a bounding box, such as an administrative boundary or a derived polygon based on time or distance, or from the *n*-nearest neighbors of the observation. Descriptive statistics can include counts, minimums and maximums, medians and means, or standard deviations.

- Interpolated spatial features, where information is borrowed and transferred from one space to another. We'll discuss this in more depth in *Chapter 11, Advanced Topics in Spatial Data Science.*

- Proximity spatial features are a measure of distance from one observation to another observation in the same dataset or a different dataset. Examples include the following:

  - The distance from an Airbnb location to the next closest Airbnb in the neighborhood. This is an example of spatial feature engineering from within the same dataset.

  - The distance from an Airbnb location to the nearest transportation station. This is an example of spatial feature engineering between two datasets.

Before we dive deeper into some hands-on coding examples, where you'll be creating summary spatial features and proximity spatial features, it is helpful to have a conversation about geospatial magic.

Performing a bit of geospatial magic

As discussed previously, spatial feature engineering can be conducted within an existing dataset where the spatial structure is already known and easy to work with. In this case, life is easy, but what about working with data from different datasets that aren't linked together?

When it comes to engineering spatial variables from two or more datasets, you must first tie the datasets together. Geospatial data is collected from varying scales and often reported at varying levels of aggregation (e.g., city, state, county, nation, and world). To complicate matters more, geospatial data isn't neat and tidy and is often stored in a variety of places in a variety of formats. Working through all of this and linking it together is where the power of geographic attributes comes in and allows you to perform geospatial magic. Geography is one of the most powerful links between things, as all things have a spatial attribute. This allows you to connect datasets together that would not typically be able to be connected, due to the lack of a primary or common key attribute.

Linking together data and normalizing its collection scale and aggregation level is a necessary step for machine learning applications. Data fed into a machine learning model needs to be logically structured and normalized as much as possible in order to achieve the most beneficial results.

You now have an understanding of the different types of spatially engineered variables; now, let's make the magic happen.

Engineering summary spatial features

In *Chapter 6, Hypothesis Testing and Spatial Randomness*, we introduced you to a dataset of store locations. This dataset contains the store locations for Dollar General, a low-price retailer that operates across the United States. The store locations were queried using the OpenStreetMap API. To walk you through how this data was queried, a supplemental notebook called `OSM POI Data Pulls` is included in the GitHub repo. This data is available under the Open Data License, and you can find out more by visiting `https://www.openstreetmap.org/copyright`. For this section, we'll continue to work with this data to begin our hands-on coding activity to create summary spatial features. Let's first import the data:

```
# Reading in the data from the path
locs_pdf = pd.read_csv(data_path + 'OSM_DollarGeneralLocs.csv')


# Converting the pandas dataframe into a geopandas geodataframe
locs_gdf = gpd.GeoDataFrame(
    locs_pdf, geometry=gpd.points_from_xy(locs_pdf.X, locs_
pdf.Y),
    crs="EPSG:4326"
)


# Resetting the index and creating a synthetic ID field
locs_gdf.reset_index(inplace=True)
locs_gdf.rename(columns={'index':'ID'}, inplace=True)
```

With the data imported, you can now start constructing some initial summary spatial features.

Summary spatial features using one dataset

Let's say that an executive at Dollar General approached you and said that he is curious whether there is a relationship between the number of stores in close proximity and the sales at each store. As a spatial data scientist, you can begin analyzing this through the use of spatial proximity features. You'll do that in the following steps:

1. Create an aggregation area around each observation:

    ```
    # To create a buffer, we first need to convert from a
    g-crs to a p-crs
    locs_gdf = locs_gdf.to_crs(3005)


    # Next, create aggregation area around each store
    ```

```
buffer_size_mi = 5
buffer_size_m = buffer_size_mi * 1609.344 # meters in a
mile

# Creating a copy of the original dataframe to operate on
locs_gdf_buffer = locs_gdf.copy()

# Performing the buffer operation
locs_gdf_buffer["buffer_5mi"] = locs_gdf.buffer(buffer_
size_m)
locs_gdf_buffer[['ID','geometry','X','Y','buffer_5mi']].
head()
```

This aggregation polygon is typically called a buffer. The buffer is created by calling the `.buffer()` method of GeoPandas. The `.buffer()` method takes as input a distance in meters. We start with a distance measure of 5 miles and convert it to meters by multiplying by 1,609.344. The output of this method is a new column appended to the copy of the store locations called `buffer_5mi`. This new column contains the geometry of a 5-mile buffer around each store location. The result of the prior code snippet is displayed in *Figure 7.2*, with the new buffer geometry appended to the DataFrame and called `buffer_5mi`.

	ID	geometry	X	Y	buffer_5mi
0	0	POINT (4662241.144 445480.319)	-82.458599	38.428581	POLYGON ((4670287.864 445480.319, 4670249.117 ...
1	1	POINT (4667682.555 450628.186)	-82.375886	38.438389	POLYGON ((4675729.275 450628.186, 4675690.528 ...
2	2	POINT (4610865.116 491258.202)	-82.633850	39.047795	POLYGON ((4618911.836 491258.202, 4618873.089 ...
3	3	POINT (4562888.418 588383.044)	-82.443563	40.034121	POLYGON ((4570935.138 588383.044, 4570896.391 ...
4	4	POINT (4519081.302 666962.782)	-82.332066	40.856058	POLYGON ((4527128.022 666962.782, 4527089.275 ...

Figure 7.2 – Buffered store locations

2. Spatially join the buffered data to the original data.

    The next step is to join the newly created buffered data to the original store location file. This will combine the two datasets based on geometry and allow you to perform an aggregation function:

```
# Joining the buffer to the store locations table
joined = gpd.sjoin(

    # Right table is the raw store locations data
    locs_gdf,
    # Left table is that of the buffers around the stores
```

```
    locs_gdf_buffer.set_geometry("buffer_5mi")[["ID",
"buffer_5mi"]],
    # The operation, or spatial predicate, you'll use is
"within"
    predicate="within"
)
```

The final step to create the summary spatial feature is to aggregate the number of stores around each observation store.

3.   Aggregate the number of stores around each observation:

```
# store count
store_count = (
    joined.groupby(
        "ID_left"
    )
    .count()
)
# Converting to a dataframe and cleaning up
store_count_df = store_count.reset_index()
store_count_df = store_count_df[['ID_left','ID_right']]
store_count_df.columns=['ID','Store_Count']
store_count_df.head()
```

The output is displayed in *Figure 7.3*.

	ID	Store_Count
0	0	2
1	1	2
2	2	1
3	3	1
4	4	1

Figure 7.3 – The store count summary spatial feature

To explore this visually, let's take a look at the buffer zone around store ID 45 in relation to all stores. In *Figure 7.4*, store ID 45 is highlighted in blue, with its 5-mile buffer zone shaded in green. All other stores are represented as red points.

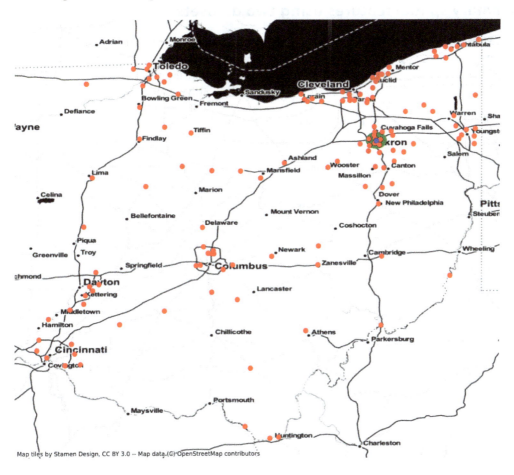

Figure 7.4 – The buffer map for store 2

To finish answering the executive's query, you would want to join information about the sales at each store and perform an analysis between the store_count variable and the sales variable. This is outside the scope of this text.

As with most processes, there are multiple ways to achieve the same result, and spatial engineering is no exception. In *Chapter 6, Hypothesis Testing and Spatial Randomness*, we introduced you to the spatial weights matrix. This matrix can be used to create proximity features that are similar to the process you just worked through in this case study.

In the next section, we'll introduce you to the next type of spatially engineered variables, which is summary spatial features.

Summary spatial features using two datasets

Let's say that the same executive has returned to you yet again and is now curious whether there is a relationship between sales and the number of competitor store locations. You can run through a similar exercise, but this time you'll use two datasets. The underlying store locations for Dollar General have not changed, and therefore, the buffers around the store have also not changed:

1. Read in the locations of Family Dollar, a competitor of Dollar General:

    ```
    # Reading in the file
    c_pdf = pd.read_csv(data_path + 'OSM_FamilyDollarLocs.
    csv')
    # Converting the pandas dataframe into a geopandas
    geodataframe
    c_gdf = gpd.GeoDataFrame(
        c_pdf, geometry=gpd.points_from_xy(c_pdf.X, c_pdf.Y),
    crs = "EPSG:4326"
    )
    # Converting to a p-CRS
    c_gdf = c_gdf.to_crs(3005)
    # Dropping records without valid geometries
    c_gdf = c_gdf[~(c_gdf['geometry'].is_empty | c_
    gdf['geometry'].isna())]
    ```

2. Perform some cleanup to create an ID and subset to include only stores in Ohio:

    ```
    # Resetting the index and creating a synthetic ID field
    c_gdf.reset_index(inplace=True)
    c_gdf.rename(columns={'index':'ID'}, inplace=True)
    # Cleaning up the data to just being those stores in Ohio
    Ohio = gpd.read_file("https://www2.census.gov/geo/tiger/
    TIGER2021/STATE/tl_2021_us_state.zip")
    Ohio = Ohio[Ohio['STUSPS']=="OH"]
    Ohio = Ohio.to_crs(3005)
    c_gdf = gpd.overlay(c_gdf, Ohio, how='intersection')
    ```

3. Spatially join the buffered data to the competitor stores:

```
# Changing to a p-crs for the buffer file
locs_gdf_buffer = locs_gdf_buffer.to_crs(3005)
# Joining the buffer to the store locations table
joined = gpd.sjoin(
    # Right table is the competitor stores
    c_gdf,
    # Left table is that of the buffers around the
primary company's Stores
    locs_gdf_buffer.set_geometry("buffer_5mi")[["ID",
"buffer_5mi"]],
    # The operation, or spatial predicate, you'll use is
"within"
    predicate="within"
)
```

4. Aggregate the number of competitor stores around each observation:

```
store_count = (
    joined.groupby(
        "ID_left"
    )
    .count()
)
# Converting to a dataframe and cleaning up
store_count_df = store_count.reset_index()
store_count_df = store_count_df[['ID_left','ID_right']]
store_count_df.columns=['ID','Comp_Store_Count']
# Displaying the data
store_count_df.head()
```

The resulting competitor store count data is displayed in *Figure 7.5*.

	ID	Comp_Store_Count
0	0	1
1	2	1
2	7	3
3	8	4
4	9	1

Figure 7.5 – Competitor store count

To complete this second analysis, you need to compare the `Comp_Store_Count` variable to a variable related to the primary company's store sales. This is beyond the scope of this text.

In this example, you definitely performed a bit of geospatial magic, as the primary company stores and competitor store locations didn't have a common feature to join them in a tabular fashion. Instead, you had to rely on geography as the common link between these two datasets. Isn't geography amazingly powerful?

Engineering proximity spatial features

In the previous section, we covered summary spatial features that are derived from preexisting spatial relationships. In this section, we'll be covering proximity spatial features that are derived based on the proximity, or distance, between two or more observations. To calculate these features, let's import the NYC Airbnb dataset that we've worked with previously. To do that, you'll run the following code cell:

1. Import the data:

```
# Reading in the data
listings = pd.read_csv(data_path + r'NY Airbnb June 2020\
listings.csv.gz', compression='gzip', low_memory=False)
# Converting it to a GeoPandas DataFrame
listings_gpdf = gpd.GeoDataFrame(
    listings,
    geometry=gpd.points_from_xy(listings['longitude'],
                                listings['latitude'],
                                crs="EPSG:4326")
)
```

For the example that you'll be working through in the next section, you'll be focusing on the borough of Manhattan in New York City. You'll need to filter the Airbnb listing dataset down to the locations that are in Manhattan.

2. Read in a shapefile of the New York boroughs and filter to only Manhattan:

```
# Focusing on attractions in Manhattan, so we need to
create a mask to filter locations in the Manhattan
borough
boroughs = gpd.read_file(data_path + r"NYC Boroughs\
nybb_22a\nybb.shp")
manhattan = boroughs[boroughs['BoroName']=='Manhattan']
manhattan = manhattan.to_crs('EPSG:4326')\
```

Next, you'll create a mask based on the Manhattan borough polygon to filter the Airbnb locations. Creating the mask allows you to perform filtering based on geographies stored in an object outside of your primary DataFrame.

3. Create a mask and filter the Airbnb data using this mask:

```
# Creating a mask
listings_mask = listings_gpdf.within(manhattan.loc[3,
'geometry'])
# Using the mask to filter the data
listings_manhattan = listings_gpdf.loc[listings_mask]
listings_manhattan.head()
```

4. Display a map to validate the results:

```
# Set up figure and axis
f, ax = plt.subplots(1, figsize=(10, 10))
# Plot all airbnb locations in green
listings_manhattan.plot(ax=ax, color="g")
# Add Stamen's Toner basemap
contextily.add_basemap(
    ax,
    crs=listings_manhattan.crs.to_string(),
    source=contextily.providers.Stamen.Watercolor
)
# Remove axes
ax.set_axis_off()
# Displaying the plot
plt.show()
```

Executing the steps outlined previously produces the map displayed in *Figure 7.6*. You can see from the map that the filtering process worked and that only the Airbnb listings in Manhattan are left in the dataset.

Figure 7.6 – The borough of Manhattan Airbnb listings

Now that the dataset is filtered, we can move on to the first exercise of creating proximity spatial features.

Proximity spatial features – NYC attractions

In this exercise, you'll be calculating some proximity spatial features to better understand the distance between some of New York's most famous attractions and the Airbnb locations in Manhattan. Let's begin:

1. First, import the NYC attractions data and convert it into a GeoPandas DataFrame:

```
# Reading in data on popular NYC Attractions
nyc_attr = pd.read_csv(data_path + 'NYC Attractions\\NYC
Attractions.csv')
# Convert PDF to GPDF
nyc_attr_gpdf =  gpd.GeoDataFrame(
    nyc_attr,
    geometry=gpd.points_from_xy(nyc_attr['Longitude'],
                                nyc_attr['Latitude'],
                                crs="EPSG:4326")

)

# Displaying the table
nyc_attr_gpdf.head(9)
```

The preceding code produces the output displayed in *Figure 7.7*. You'll see nine of the most famous attractions listed, which include the Empire State Building, Central Park, and Times Square.

	Attraction	Latitude	Longitude	geometry
0	Central Park	40.7851	-73.9683	POINT (-73.96830 40.78510)
1	Central Park Zoo	40.7678	-73.9718	POINT (-73.97180 40.76780)
2	Empire State Building	40.7484	-73.9857	POINT (-73.98570 40.74840)
3	Statue of Liberty	40.6892	-74.0445	POINT (-74.04450 40.68920)
4	Rockeffeller Center	40.7587	-73.9787	POINT (-73.97870 40.75870)
5	Chrysler Building	40.7516	-73.9755	POINT (-73.97550 40.75160)
6	Times Square	40.7580	-73.9855	POINT (-73.98550 40.75800)
7	MoMa	40.7614	-73.9776	POINT (-73.97760 40.76140)
8	Charging Bull	40.7046	-74.0139	POINT (-74.01390 40.70460)

Figure 7.7 – NYC attractions

2. Plot the data on a map to understand the locations better:

```
# Set up figure and axis
f, ax = plt.subplots(1, figsize=(10, 10))
# Plot all attractions in blue
nyc_attr_gpdf.plot(ax=ax, color="b")
# Add Stamen's Toner basemap
contextily.add_basemap(
```

```
        ax, crs=nyc_attr_gpdf.crs.to_string(),
        source=contextily.providers.Stamen.Watercolor
)
# Remove axes
ax.set_axis_off()
# Display the plot
plt.show()
```

Figure 7.8 shows the nine New York city attractions on a map oriented around the borough of Manhattan.

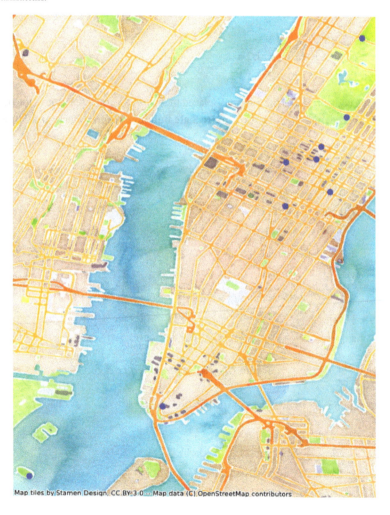

Figure 7.8 – The NYC attractions map

With the locations of the NYC attractions now ready, let's overlay them onto the map of the Airbnb listings. You'll do that in two parts.

3. Set up the initial map:

```
from matplotlib.lines import Line2D
# Set up figure and axis
f, ax = plt.subplots(1, figsize=(10, 10))
# Plot all airbnb locations in green
listings_manhattan.plot(ax=ax, color="g")
# Plot all attractions in blue
nyc_attr_gpdf.plot(ax=ax, color="b")
# Add Stamen's Toner basemap
contextily.add_basemap(
    ax, crs=nyc_attr_gpdf.crs.to_string(),
    source=contextily.providers.Stamen.Watercolor
)
```

4. Add a legend and remove the axis to produce a clean visualization:

```
# Remove axes
ax.set_axis_off()
# Manually creating a legend to orient audience
green_circle = Line2D([0], [0], marker='o', color='w',
label='Airbnbs', markerfacecolor='g', markersize=8)
blue_circle = Line2D([0], [0], marker='o', color='w',
label='Attractions', markerfacecolor='b', markersize=8)
plt.legend(handles=[green_circle, blue_circle])
# Display the map
plt.show()
```

Figure 7.9 shows the output plot of the preceding code block:

Figure 7.9 – The Manhattan Airbnb listings and NYC attractions

5. Now, let's calculate the distance between the Manhattan Airbnb listings and the NYC attractions:

```
# Calculate the distance to each attraction per Airbnb
attractions = nyc_attr_gpdf.Attraction.unique()
# Converting to a projected coordinate system
nyc_attr_gpdf_p = nyc_attr_gpdf.to_crs('EPSG:2263')
listings_manhattan_p = listings_manhattan.to_
crs('EPSG:2263')
```

```
# Applying a lambda function that calls geopandas
distance function to calculate the distance between each
Airbnb and each attraction
distances = listings_manhattan_p.geometry.apply(lambda g:
nyc_attr_gpdf_p.distance(g)).head()
# Renaming the columns based on the attraction for which
the distance is calculated
distances.columns = attractions
# Displaying the top 5 rows of the dataframe
distances.head()
```

The resulting distances are displayed in *Figure 7.10*. You can see that the 0th indexed Airbnb location is 12,449.13 units away from Central Park. Not knowing the units of measurement isn't beneficial, so let's figure that out.

	Central Park	Central Park Zoo	Empire State Building	Statue of Liberty	Rockeffeller Center
0	12449.142142	6442.861743	1880.201525	28573.983993	2674.064858
3	8538.692949	3362.221500	5932.808083	32295.350361	2471.262755
4	6816.568517	13169.722567	20803.328087	46886.699521	16721.142383
5	6693.240013	13068.682235	20846.937438	47148.706760	16710.851616
8	9202.777723	14578.797153	22606.799884	49590.742578	18399.171063

Figure 7.10 – The distance between Airbnbs and NYC attractions

6. Identify the unit of measurement for the projected coordinate reference system:

```
# To understand what the distance unit is, we run the
following function
listings_manhattan_p.crs.axis_info[0].unit_name
```

Running the previous function informed you that the unit of measurement is in the US survey foot or, more simply, feet. Distance measurements in feet aren't great for a big city where you'll likely be traveling by foot or the subway.

7. Convert the measurement into the more commonly used miles unit of measurement:

```
# Convert from 'US survey foot' to miles
distances = distances.apply(lambda x: x/5280, axis=1)
distances.head()
```

The output of running the prior code cell is the DataFrame displayed in *Figure 7.11*. From this, we can now see that the 0th indexed Airbnb is located approximately 2.35 miles away from Central Park.

	Central Park	Central Park Zoo	Empire State Building	Statue of Liberty	Rockeffeller Center	Chrysler Building	Times Square	MoMa	Charging Bull
0	2.357790	1.220239	0.356098	5.411749	0.506452	0.546463	0.306406	0.684410	3.690749
3	1.617176	0.636786	1.123636	6.116553	0.468041	0.981280	0.469539	0.365030	4.441309
4	1.291014	2.494261	3.940019	8.880064	3.166878	3.626230	3.298179	2.973187	7.264976
5	1.267657	2.475125	3.948279	8.929687	3.164929	3.609690	3.316261	2.970171	7.289971
8	1.742952	2.761136	4.281590	9.392196	3.484691	3.826188	3.729309	3.297390	7.637321

Figure 7.11 – The distances between Manhattan Airbnb locations and NYC attractions

The prior steps demonstrated how to calculate straight-line proximity spatial features. However, we can take it one step further by aggregating the raw distances into additional distance bands that summarize the number of attractions within 1 to 6 miles for every Airbnb.

8. Create summary features by distance bands:

```
# Check to see which locations are less than 2 miles
distances_1mi = distances.apply(lambda x: x <=1, axis=1).
sum(axis=1)
distances_2mi = distances.apply(lambda x: x <=2, axis=1).
sum(axis=1)
distances_3mi = distances.apply(lambda x: x <=3, axis=1).
sum(axis=1)
distances_4mi = distances.apply(lambda x: x <=4, axis=1).
sum(axis=1)
distances_5mi = distances.apply(lambda x: x <=5, axis=1).
sum(axis=1)
distances_6mi = distances.apply(lambda x: x <=6, axis=1).
sum(axis=1)
# Creating a dataframe combining all the distance bands
distance_df = pd.concat([distances_1mi,distances_2mi,dis-
tances_3mi,distances_4mi,distances_5mi,distances_6mi],
axis=1)
distance_df.columns =
['Attr_1mi','Attr_2mi','Attr_3mi','Attr_4mi','At-
tr_5mi','Attr_6mi']
distance_df.head()
```

The output of the prior code cell is displayed in *Figure 7.12*.

	Attr_1mi	Attr_2mi	Attr_3mi	Attr_4mi	Attr_5mi	Attr_6mi
0	5	6	7	8	8	9
3	5	7	7	7	8	8
4	0	1	3	7	7	7
5	0	1	3	7	7	7
8	0	1	2	6	7	7

Figure 7.12 – A count of attractions

To wrap things up, you need to merge the distance data back together with the original dataset. By combining the data with the original data, you'll have a good starting point when you want to conduct downstream processes such as modeling.

9. Join the distances and the summary distance bands to the original dataset:

```
# Joining back to the listings geopandas df
listings_manhattan = listings_manhattan.merge(distances,
left_index=True, right_index=True)
listings_manhattan = listings_manhattan.merge(distance_
df, left_index=True, right_index=True)
```

In this section, you've learned how to create proximity-based spatial features by calculating the straight-line distance between the Airbnbs in Manhattan and popular tourist attractions. These proximity features will come in handy in *Chapter 9, Developing Spatial Regression Models*.

Summary

In this chapter, we introduced you to the concept of spatial feature engineering. Recall that spatial feature engineering falls into two classes: summary spatial features and proximity spatial features. These two classes are respectively based on summarizing spatial data based on preexisting spatial relationships, and the distance, or proximity, between observations.

During the chapter, we performed two exercises based on data pertaining to Dollar General stores and its competitor, Family Dollar, and also based on Manhattan Airbnb locations and nearby NYC attractions. Throughout these exercises, we leveraged concepts we introduced you to in previous chapters, such as filtering based on masks, converting pandas DataFrames into GeoDataFrames, and working with projected coordinate reference systems.

Finally, we went over the concept of *geospatial magic* and the power that geography has as a universal link between data and objects. We hope that you are beginning to see the power that geospatial data and methods will bring to your analysis in the future.

This concludes *Part 3, Exploratory Spatial Data Analysis. Chapter 8, Spatial Clustering and Regionalization,* will kick off the final part of this book, which is *Part 4, Geospatial Modeling Case Studies.* Everything that you've learned up until now will enable you to build some truly powerful models in the next section.

Part 3:
Geospatial Modeling
Case Studies

In Part 3 of this book, you'll be exposed to a number of different modeling topics and apply those directly to several case studies. The modeling topics focus on clustering, regionalization, regression, and optimization problem types. Throughout this section, you'll be provided with hands-on coding exercises to solidify your learning. These coding exercises can be leveraged throughout your journey as a spatial data scientist, as they can be adapted and expounded upon to solve problems across industries. The last chapter of this book focuses on more advanced topics such as spatial interpolation, spatial indexing, and ethics in spatial data science.

This part comprises the following chapters:

- *Chapter 8, Spatial Clustering and Regionalization*
- *Chapter 9, Developing Spatial Regression Models*
- *Chapter 10, Developing Solutions for Spatial Optimization Problems*
- *Chapter 11, Advanced Topics in Spatial Data Science*

8

Spatial Clustering and Regionalization

You've learned a lot so far and it has brought you to the final section of this book, *Part 3, Geospatial Modeling Case Studies*. In this section, you'll leverage all of the skills you gained so far and develop additional skills, as you work to implement geospatial models throughout a number of case study exercises. These case studies are applicable across a number of industries and will provide you with code that can be modified and enhanced in your work down the road.

In this chapter, we will discuss how you can use geospatial data and methods to assemble your observations into groups known as **clusters**. The process of creating clusters is known as **clustering**, which leverages **unsupervised machine-learning** techniques to define the clusters. Clustering is known as an unsupervised process because there is no ground truth value that you're training your algorithm on. Instead, clustering attempts to derive structure from the underlying data. In *Chapter 9, Developing Spatial Regression Models*, we'll be discussing regression-based methods that are known as **supervised machine learning** because there is a target value that we'll be trying to predict or explain.

As a spatial data scientist, you can leverage clustering to help you better understand the similarities within and the differences between each cluster. For instance, you may find that certain clusters have higher rates across some variables while consistently lower rates across others. Understanding these similarities and differences will help you derive meaningful insights from complex data derived from a real-world process that you're looking to better understand. These insights can be distilled down into **cluster profiles**, which can be used to easily describe each group.

Within this chapter, we'll review a variety of clustering algorithms that leverage spatial data and spatial constraints in their development. Once the clusters are developed, we'll evaluate the clusters through a process known as **profiling** to derive cluster profiles. We'll do this through the use of geographic visualizations and descriptive statistics. In the last section of this chapter, we'll introduce you to ways you can evaluate your clustering algorithm's performance using common mathematical measures.

In this chapter, you'll learn the following:

- How to pull geodemographic data from the US Census Bureau API

- How to implement clustering or segmentation models

- How to implement regionalization models

- How to profile your segmentation models

- How to measure the performance of your model

Technical requirements

For this chapter, you'll leverage the Jupyter notebook called `Chapter 8 - Spatial Clustering NY`, which is stored in the GitHub repo for this book at `https://github.com/PacktPublishing/Applied-Geospatial-Data-Science-with-Python/tree/main/Chapter08`.

Collecting geodemographic data for modeling

Before you start developing models, it is critical that you gather, clean, explore, and process data in a way that will lead to the most effective clustering models. You may recall that these four steps are the first four steps in the data science pipeline we've discussed throughout this book. To begin, you'll leverage the Census API to collect geodemographic data.

Extracting data using the Census API

The clustering exercise that you'll work through later on in this chapter focuses on building out geodemographic clusters for **New York City** (**NYC**). To do this, you'll first need to collect data utilizing the US Census Bureau API. To pull data via this API, you'll need to request an API key by visiting `https://api.census.gov/data/key_signup.html`. Requesting an API key and pulling data from the Census Bureau is free and open to the public. After requesting a key, you will be given a unique 40-digit alphanumeric string that is unique to you. Keep track of this key and ensure it is stored in a safe place.

With your API key in hand, you can now begin setting up your notebook to pull data. We encourage you to follow along using the steps defined in the companion with the code for this chapter included in the Jupyter notebook called `Chapter 8 - Spatial Clustering`.

1. Import the required packages using the following code:

```
from census import Census
from us import states
```

Here, you'll notice the two lines of the code refer to two packages that you have not yet worked with. The `census` package allows you to easily access data from the US Census Bureau using their API. The `us` package is an easy-to-use package for working with US and state-level metadata, which includes information, such as their FIPS codes and URLs to their shapefiles. As a reminder, **FIPS** stands for **Federal Information Processing Code** and was introduced to you back in *Chapter 2, What Is Geospatial Data and Where Can I Find It?*, along with other Census Bureau metadata.

2. Next, we'll load your Census API key. You'll need to pass the API key that you were given to the `census` package. You'll want to ensure that your API key is wrapped with quotes, as seen in the next code block. This ensures that it is passed as a string to the API, which is an API-level requirement:

```
c = Census("CENSUS API KEY HERE")
```

Using the Census API, you'll be pulling census data at the tract level for New York. The census package provides you with convenient methods for pulling data from a wide array of Census administrative boundary geographies. To refresh your memory on Census geographies, you can refer to the *Human geography* section of *Chapter 2, What Is Geospatial Data and Where Can I Find It?*. The data that you'll be pulling will come from the **Annual Community Survey** (**ACS**). To find more information about the ACS and the variables within it, you can explore the data dictionary by visiting `https://api.census.gov/data/2019/acs/acs5/variables.html`.

3. We'll create a list of ACS variables to be used for clustering. We've predefined a list of variables that relate to the vitality of an area with information pertaining to income, population, education, and the number of people living below the poverty line. You'll pass those variables to a list that you'll access throughout the case study. A truncated version of the code block is included for reference:

```
geo_demo = [
    "B01003_001E", #"Total Population"
    "B25077_001E", #"Median value of owner-occupied
units"
    "B25026_001E", #"Total population in occupied housing
units",
    ...
    ]
```

4. Then, we'll pass the selected variables to the Census API. The Census API takes a few key inputs:

    - `fields`: The variables you're interested in
    - `State_fips`: The state you're requesting data for
    - `county_fips: "*"`: For all counties in the given state or a specific county FIPS code
    - `tract: "*"`: For all tracts in the given state or a specific tract FIPS code
    - `year`: The year of interest

    To better understand the Census API, visit the source documentation here: `https://www.census.gov/content/dam/Census/data/developers/api-user-guide/api-guide.pdf`.

    The next code block shows how these parameters are passed to the API. The code is truncated where the ellipses are for brevity:

    ```
    ny_census = c.acs5.state_county_tract(fields = ('NAME',
    'B01003_001E','B25026_001E','B25008_002E','B25008_003E',
    ...
                                                'B060
    10_011E','B28007_009E','B19059_002E','B19059_0
    03E',                                        'B08-
    013_001E','B17013_002E'),
                                            state_fips =
    states.NY.fips,
                                            county_fips = "*",
                                            tract = "*",
                                            year = 2019)
    ```

    The Census API returns data via a pandas DataFrame. In order to perform spatial operations, you'll need to geo-enable the data by joining the geometry attributes to it.

5. Geo-enable the geodemographic data. To geo-enable the data, you'll first pull a shapefile of the NY census tracts. To pull the census tract, you'll need to identify the two-digit State FIPS code for NY. You can explore a list of State FIPS codes by visiting `https://www.census.gov/library/reference/code-lists/ansi.html`, and scrolling to the section titled *FIPS Codes for the States and the District of Columbia*. The two-digit FIPS code for New York is 36. Next, you'll need to determine the best projection for New York data. `Spatialreference.org` is a great site to visit to learn more about the best spatial reference for various geographies. For this exercise, you'll reproject the shapefile to EPSG:2263, the New York Long Island Plane. The New York Long Island Plane projection is a projection that will have the least amount of skew and is a better-suited projection for data concentrated within New York.

```
# Access shapefile of NY census tracts
ny_tract = gpd.read_file("https://www2.census.gov/geo/
tiger/TIGER2019/TRACT/tl_2019_36_tract.zip")
# Reprojecting the shapefile to the New York State Plan
Long Island Zone EPSG:2263 - https://spatialreference.
org/ref/epsg/2263/
ny_tract = ny_tract.to_crs(epsg = 2263)
# Print GeoDataFrame of the NY census tract shapefile
print(ny_tract.head(2))
print('NY Tract Shape: ', ny_tract.shape)
# Check the projection of the shapefile
print("\nThe shapefile projection for this data is: {}".
format(ny_tract.crs))
```

6. Next, you'll perform some variable manipulation to create a new variable called GEOID, which will represent the tract FIPS code for each observation of the ny_df DataFrame:

```
# Combine the state, county, and tract variables of the
ny_df together to create a new string and assign to
variable called GEOID
ny_df["GEOID"] = ny_df["state"] + ny_df["county"] + ny_
df["tract"]
# Remove the individual columns as they're no longer
needed
ny_df = ny_df.drop(columns = ["state", "county",
"tract"])
```

7. Finally, you'll merge the Census data stored in ny_df with the geographic data stored in the ny_tract data, which will geo-enable the census data. You'll do that in the following code block:

```
# Join the data together on GEOID to geoenable the census
data
ny_merge = ny_tract.merge(ny_df, on = "GEOID")
# Display the results
ny_merge.head(2)
```

Figure 8.1 displays the resulting DataFrame:

	STATEFP	COUNTYFP	TRACTCE	GEOID	NAME_x	NAMELSAD	MTFCC	FUNCSTAT	ALAND	AWATER	...	B06010_007E	B06010_008E
0	36	081	044800	36081044800	448	Census Tract 448	G5020	S	208002	0	...	234.0	104.0
1	36	081	045800	36081045800	458	Census Tract 458	G5020	S	245281	0	...	187.0	167.0

Figure 8.1 – NY tracts with census data

With the data pulled, you can now move on to cleaning it. You'll perform several cleaning tasks in the next section.

Cleaning the extracted data

One of the first cleaning steps that will assist you later on in the exercise is renaming the variables from the Census Bureau's naming conventions to ones that are a bit more interpretable without referring to the data dictionary. You'll do that by calling the .rename() method on the DataFrame, as displayed in the next code block. The code is truncated for brevity.

1. Let's rename the variables to make them easily interpretable:

```
# Renaming variables in the data set for interpretability
ny_merge.rename(columns={
    "B01003_001E":"TotPop", #"Total Population"
    "B25077_001E":"MedVal_OwnOccUnit", #"Median value of
owner occupied units"
    ...
    "B08013_001E":"TrvTimWrk", #"Travel time to work in
minutes"
    "B17013_002E":"PopBlwPovLvl" #"Population with income
below poverty level in past 12 months"
    }
                , inplace=True
            )
```

Next, you'll create a list of variables that will be used to subset the DataFrame to only the variables that will be used in the clustering exercise later on. The new DataFrame will be called ny_merge_2.

2. Now, we'll subset the DataFrame:

```
geo_demo_rn = [
"TotPop", #"Total Population"
"TotPopOccUnits", #"Total population in occupied housing
units"
...
"RetPopNoRetInc", #"Retired without retirement income"
"PopBlwPovLvl" #"Population with income below poverty
level in past 12 months"
]
# Cleaning up the dataframe
geo_demo_rn.append('geometry')
ny_merge_2 = ny_merge[geo_demo_rn]
geo_demo_rn.remove('geometry')
```

Lastly, you'll perform a filtering operation to remove any census tracts where there is no population, as these areas will not benefit your clustering analysis.

3. Finally, remove areas without a population:

```
# Dropping any areas without population
ny_merge_2 = ny_merge_2[ny_merge_2['TotPop']>0]
# Resetting the index to assist in index-based operations
later on
ny_merge_2.reset_index(inplace=True)
```

Now that you have a clean dataset, you can begin the exploratory data analysis phase, which will include both traditional **exploratory data analysis (EDA)** and its spatial relative, **exploratory spatial data analysis (ESDA)**. As a reminder, we covered these topics in detail in *Chapter 5, Exploratory Data Visualization,* and *Chapter 6, Hypothesis Testing and Spatial Randomness.*

Conducting EDA and ESDA

You'll start by visualizing each of the variables that you extracted from the Census API:

1. Set up the figure using subplots to map each extracted variable from the Census API:

```
# Setting up the figure using subplots to map each of the
extracted variables
fig, axes = plt.subplots(nrows=7, ncols=3,
figsize=(75,75), layout='tight')
```

```
axes = axes.flatten()

# Setting the font size
plt.rcParams['font.size'] = '40'
```

2.   Iterate through the list of variables and plot each variable using quantiles:

```
# Plotting each of the extracted variables in a subplot
for ind, col in enumerate(geo_demo_rn):
    ax = axes[ind]
    ny_merge_2.plot(column=col,
                    ax = ax, scheme = "quantiles",
linewidth=0, cmap="coolwarm",
                    legend=True, legend_kwds={'loc':
'center left','bbox_to_anchor':(1,0.5),'fmt': "{:.0f}"}
                    )
    ax.set_axis_off()
    ax.set_title(col)
plt.subplots_adjust(wspace=None, hspace=None)
plt.show()
```

The first four rows of the subplots resulting from the prior code block are displayed in *Figure 8.2*:

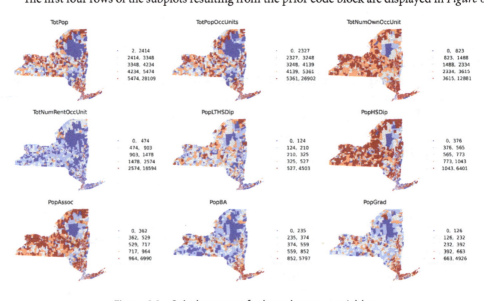

Figure 8.2 – Subplot maps of selected census variables

From the preceding subplots, you can see areas with higher values displayed in red, and areas with lower values displayed in blue. Visually, there are a number of patterns that begin to emerge. First, you can see a higher proportion of the population with an associate's degree or less education. When you look closer at Long Island and the NYC boroughs, there are larger numbers of the population with bachelors and graduate-level degrees. The total number of renter-occupied units is also very low in the majority of the state with the exception of the large numbers of rental units in the NYC boroughs.

The differences and commonalities across space will benefit the performance of your clustering models and the differentiation of your resulting cluster profiles. If all of the spatial visuals showed relatively similar patterns, then there would not be a huge benefit gained from leveraging spatial clustering methodologies.

Measuring spatial autocorrelation

Next, we'll explore the degree of spatial autocorrelation present in the data. In *Chapter 6*, *Hypothesis Testing and Spatial Randomness*, we introduced you to Moran's I as one of the more common measures of global spatial autocorrelation. In the next code block, you'll use Moran's I to formally test for the presence of a non-random spatial distribution in the data:

1. First, you'll need to establish the spatial weights matrix:

    ```
    # Importing packages required for testing spatial
    autocorrelation
    from libpysal.weights import Queen, KNN
    from esda.moran import Moran
    import numpy as np
    # Testing for spatial auto correlation using Moran's I.
    First, we need to set up the spatial weights matrix
    w = Queen.from_dataframe(ny_merge_2)
    ```

    With the spatial weights matrix defined, you can now calculate the Moran's I statistic for each variable. In the next code block, you'll start by setting a random seed in NumPy. This random seed will ensure the reproducibility of your code, which is important to ensure you get the same results each time you run the code:

    ```
    # Set the numpy random seed for reproducibility
    np.random.seed(54321)


    # Calculate the Moran's I statistic for each
    geodemographic variable
    moransi_results = [
        Moran(ny_merge_2[v], w) for v in geo_demo_rn
    ]
    ```

```
# Structure results as a list of tuples
moransi_results = [
    (v, res.I, res.p_sim)
    for v, res in zip(geo_demo_rn, moransi_results)
]
# Display as a table
table = pd.DataFrame(
    moransi_results, columns=[" GEODEMO Var", "Moran's
I", "P-value"]
).set_index("GEODEMO Var")
table.head(5)
```

The Moran's I statistic and p-value for the first five geodemographic variables are displayed in
Figure 8.3:

GEODEMO Var	Moran's I	P-value
TotPop	0.293936	0.001
TotPopOccUnits	0.285091	0.001
TotNumOwnOccUnit	0.577288	0.001
TotNumRentOccUnit	0.544159	0.001
PopLTHSDip	0.504985	0.001

Figure 8.3 – Moran's I statistic for selected variables

For the selected variable, as well as those not displayed, there is a statistically significant non-random
spatial pattern inherent in the data.

Another way to explore the data is by producing pairplots using the seaborn package that displays
the correlation between variables. Given that numerous variables have been pulled from the Census
API, there would be far too many plots to visually inspect. For this step of the exercise, you'll only
produce pairplots for a few selected variables. You'll first import seaborn as sns, and then plot the
pairplots for the selected variables:

```
# Importing packages required for additional visualization
import seaborn as sns


# Given we have 25 variables in the data set to explore, this
will be way to many plots to visually inspect. Lets inspect
just a handful
sel_vars = ["TotPop", #"Total Population"
```

```
"TotNumRentOccUnit", #"Total number of renter occupied units",
...,
"PopBlwPovLvl" #"Population with income below poverty level in
past 12 months"
]
pplt = sns.pairplot(
    ny_merge_2[sel_vars], kind="reg", diag_kind="kde"
)


plt.show()
```

The resulting pairplots are displayed in *Figure 8.4*. From the following screenshot, you can see a handful of patterns emerge. One of the most important and intuitive patterns is the negative correlation between the population with a graduate degree and the population below the poverty line. These data correlations will benefit the clustering algorithm, as this information will allow it to separate higher and lower-educated areas of the state into distinct clusters.

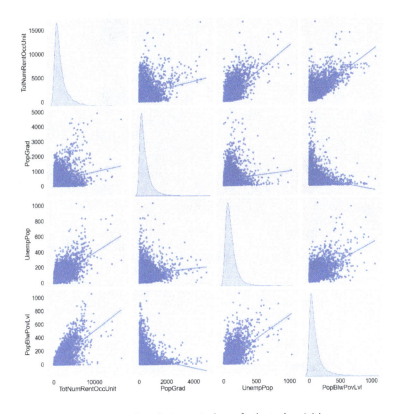

Figure 8.4 – Correlation pairplots of selected variables

Another way to plot the pairwise correlations between the geodemographic variables is by producing a correlation heatmap. The following code block demonstrates how to do this:

```
# Set the figure size
plt.figure(figsize=(16, 12))
# Setting the font size
plt.rcParams['font.size'] = '10'
# Create the mask to only show the lower triangle
mask = np.triu(np.ones_like(ny_merge_2.corr(), dtype=np.bool))
heatmap = sns.heatmap(ny_merge_2.corr(), mask=mask, vmin=-1,
vmax=1, annot=True, cmap='coolwarm')
heatmap.set_title('ACS Variable Correlation Heatmap',
fontdict={'fontsize':18}, pad=16);
```

Figure 8.5 displays the resulting pairwise correlation heatmap:

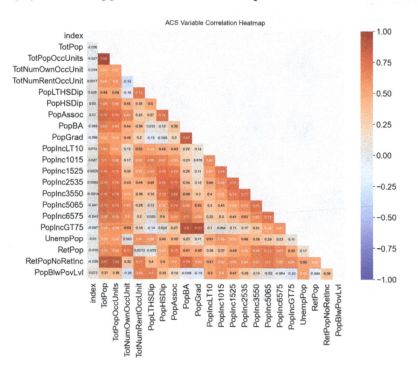

Figure 8.5 – Correlation heatmap for all variables

From the correlation heatmap, you can see strong correlations between higher education levels and higher income levels above $65,000. There is also a very low correlation of 0.11 between the unemployed population and the population with a graduate-level degree. The clustering algorithm you implement later on in this chapter will pick up on these signals and others to create differentiated groupings.

Now that you've explored the data, it is important that you standardize the data, as the computer does not know how to interpret the scales of each variable. To do this, you'll import the `robust_scale` module from `sklearn.preprocessing`. With the `robust_scale` module, you'll create a new scaled dataset, as shown in the next code block:

```
# Importing the packages needed to scale the data
from sklearn.preprocessing import robust_scale

ny_merged_scaled = robust_scale(ny_merge_2[geo_demo_rn])
ny_merged_scaled
```

Scaling the dataset is the last preprocessing step that needs to be conducted before modeling can commence. In the next section, you'll start to learn about various types of clustering models, beginning with K-means.

Developing geodemographic clusters

To begin your clustering exercise, it is helpful to talk about a few types of clustering algorithms that you'll be leveraging within this section. Your first model will be developed using a **K-means clustering algorithm**. The K-means clustering algorithm aims to split your observations into a predefined number of clusters that minimizes **within-cluster variance**. Within-cluster variance measures the similarity of observations that are grouped together in the same cluster. Later on in this section, we'll discuss how to develop clustering models using an **agglomerative hierarchical clustering (AHC) algorithm**. Agglomerative clustering begins with each observation in its own cluster and recursively merges pairs of clusters together based on a linkage metric. Similar to K-means, AHC aims to minimize within-cluster variance while maximizing between-cluster variance. There are many other types of clustering algorithms out there, such as **density-based spatial clustering of applications with noise (DBSCAN)**, **affinity propagation**, and **spectral**. These clustering methods are outside the scope of this book. To learn more about these clustering algorithms you can read Machine Learning Mastery's article entitled *10 Clustering Algorithms with Python* by visiting `https://machinelearningmastery.com/clustering-algorithms-with-python/`. You can also review the documentation for the hdbscan package called *Comparing Python Clustering Algorithms* at `https://hdbscan.readthedocs.io/en/latest/comparing_clustering_algorithms.html`.

K-means geodemographic clustering

Let's start by building out a K-means model utilizing the scaled NY geodemographic dataset. This model will act as a baseline model, which we will compare other models to. To start, you'll import the K-means module from `sklearn.cluster`. You'll also set a random seed for reproducibility:

```
# Exploring an initial k-means baseline model
from sklearn.cluster import KMeans
# setting the random seed to ensure reproducibility
np.random.seed(54321)
```

As we mentioned in the opening paragraph, a predefined number of clusters must be passed to the K-means algorithm. In order to find the optimal number of clusters, you typically look at what is known as the **elbow plot**. The elbow plot is created by plotting the **distortion** against the number of clusters, k. Distortion is defined as the average of the squared distances from each observation to the centroid, or center, of the cluster. Let's now create an elbow plot based on our data by simulating models with 1 to 15 clusters and calculating the distortion:

```
distortions = []
K = range(1,15)
for k in K:
    # Instantiating the model
    kmeans=KMeans(n_clusters=k)
    kmeans.fit(ny_merged_scaled)
    distortions.append(kmeans.inertia_)
plt.figure(figsize=(16,10))
plt.plot(K, distortions, 'bx-')
plt.xlabel('k')
plt.ylabel('Distortion')
plt.title('Elbow Method for optimal k')
plt.show()
```

The resulting elbow plot is displayed in *Figure 8.6*:

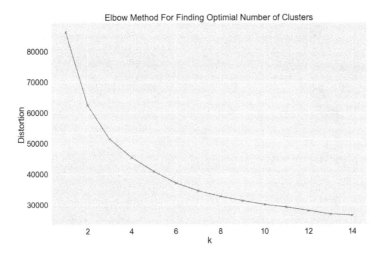

Figure 8.6 – Elbow plot

The optimal number of clusters is determined by finding the bend in the elbow of the plot or the point at which the steepness of the curve begins to decline. Here that point looks to be around five clusters. With the optimal number of clusters identified, you can now create a k-means model with five clusters:

```
# Running the KMeans model with 5 clusters
kmeans=KMeans(n_clusters=5)
kmeans_5 = kmeans.fit(ny_merged_scaled)
# Printing the cluster labels
kmeans_5.labels_
```

With the model built, you can create a choropleth map to see how the resulting clusters distribute across the state of New York. To do that, you'll execute the next code block:

```
# Assign labels to a new column called km_5_label
ny_merge_2["kmeans_5_label"] = kmeans_5.labels_
# Setup figure and axis
f, ax = plt.subplots(1, figsize=(6, 6))
# Plot the choropleth map
ny_merge_2.plot(
    column="kmeans_5_label", categorical=True,
    legend=True, linewidth=0, ax=ax,
```

```
    legend_kwds={'loc': 'center left',
                'bbox_to_anchor':(1,0.5),'fmt': "{:.0f}"},
    cmap='Set2'
)
# Remove the axis
ax.set_axis_off()
# Display the map
plt.show()
```

The code produces the visual shown in *Figure 8.7*:

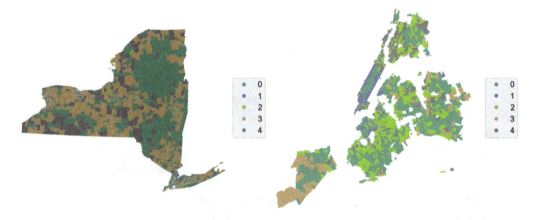

Figure 8.7 – K-means geodemographic choropleth map of New York

The map shows a few concentrations of clusters across space. Cluster **0** has a heavy concentration in the northern part of New York around the Adirondacks and Lake George area. Cluster **1** is almost not visible in the full state map as it is more centralized in NYC and the boroughs. The zoomed-in portion of the map focuses on this area.

To better understand how the clusters differ from one another, you'll typically perform what is known as **cluster profiling**. This is the process of calculating descriptive statistics to describe the differences and similarities of each cluster.

First, you'll calculate the number of census tracts that fall within each of the clusters by executing the next code block:

```
# Group data table by cluster label and count observations
k5distr = ny_merge_2.groupby("kmeans_5_label").size()
k5distr
```

The preceding code produces the table displayed in *Figure 8.8*:

Cluster Label	Number of Tracts
0	2080
1	180
2	266
3	1229
4	1099

Figure 8.8 – K-means cluster tract count

Next, you can calculate the average area of the tracts that fall within each of the clusters. You'll then plot the results in a bar plot:

```
# Getting the unit of measurement from the CRS
print(ny_merge_2.crs.axis_info)
# Getting the unit of measurement from the CRS
print(ny_merge_2.crs.axis_info)
# Calculate the average area of each cluster
# 1. Create a new column with the area of the census tract and
convert from foot to sq. mi.
ny_merge_2['area'] = (ny_merge_2.geometry.area)*3.587E-8
# 2. Dissolve the tracts and calculate the area
area = ny_merge_2.dissolve(by="kmeans_5_label", aggfunc="sum")
["area"]
print("\nThe area of the clusters is: {}".format(area))
# 3. Create a table with the number of tracts per cluster and
the sum area
tract_area = pd.DataFrame({"Num. Tracts": k5distr, "Area":
area})
tract_area['Area_per_tract'] = tract_area["Area"]/tract_
area["Num. Tracts"]
tract_area.reset_index(inplace=True)
# 4. Plot the area per tract
ax = tract_area.plot.bar(x="kmeans_5_label",y="Area_per_tract")
```

The resulting bar plot is displayed in *Figure 8.9* and shows that clusters **1**, **2**, and **4** have the smallest average tract area:

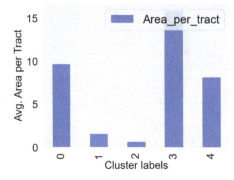

Figure 8.9 – Average tract area per cluster

Now that you've explored how the clusters differentiate in terms of geographies and area, let's look at how their statistics differ from one another based on the geodemographic variables used to create the model. To do this, you'll leverage the scaled dataset and plot a radial or spider plot. You'll perform this process by running the next code block:

```
# Creating a dataframe version of the scaled data
ny_merged_scaled_df = pd.DataFrame(ny_merged_scaled,
                  columns = geo_demo_rn)
# Adding in the cluster labels
ny_merged_scaled_df["km_6_label"] = km_6.labels_
# Calculating descriptive statistics for each cluster
k6means_s = ny_merged_scaled_df.groupby("km_6_label")[geo_demo_
rn].mean()
# Transpose the table and rounding the values to 2 decimal
places
k6means_s.round(2)
import plotly.graph_objects as go
categories = k6means_s.columns
fig = go.Figure()
for g in k6means_s.index:
    fig.add_trace(go.Scatterpolar(
        r = k6means_s.loc[g].values,
        theta = categories,
        fill = 'toself',
```

```
              name = f'cluster #{g}'
        ))
  fig.update_layout(
    polar=dict(
      radialaxis=dict(
        visible=True,
        range=[-2, 5] # here we can define the range
      )),
    showlegend=True,
      title="Cluster Radial Plot",
      title_x=0.5
  )
```

By executing the prior code cell, you'll have produced the radial plot displayed in *Figure 8.10*:

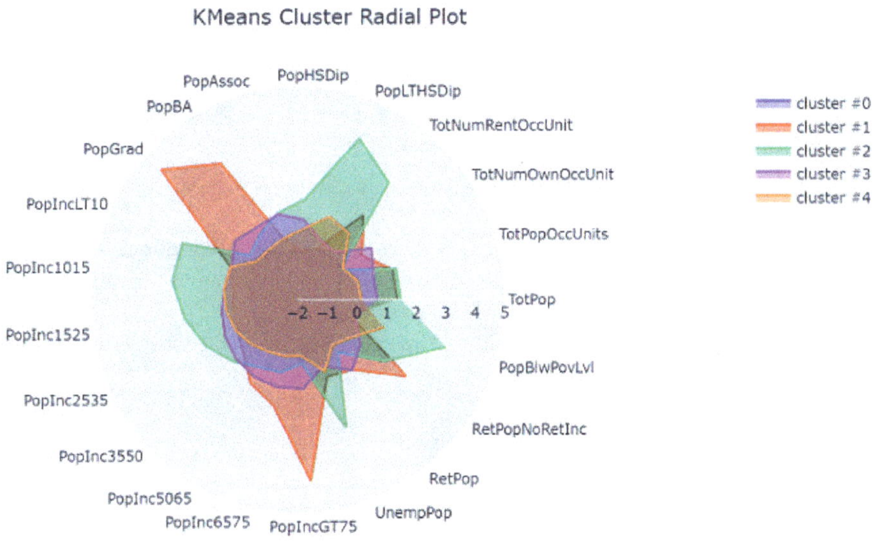

Figure 8.10 – Cluster radial plot

Because this is plotted using the scaled version of the data, the scale of the plot is not interpretable, but on a relative basis, you can tell some key differences between the clusters. For example, **cluster #1** has the highest population with a graduate and bachelor's level degree and also the highest population with an income above $75,000.

In the next section, we'll discuss clustering through the use of an AHC algorithm.

Agglomerative hierarchical geodemographic clustering

As we discussed previously, AHC starts with each observation in its own cluster and then interactively combines them into a specified number of clusters. For this exercise, we'll continue to use five clusters. First, you'll import the `AgglomerativeClustering` module from `sklearn.cluster` and again set a random seed for reproducibility:

```
# Importing the package needed for hierarchical clustering
from sklearn.cluster import AgglomerativeClustering
# Set seed for reproducibility
np.random.seed(54321)
```

Next, you'll instantiate the algorithm and store the labels inside the scaled and unscaled dataset:

```
# Instantiate the algorithm
model = AgglomerativeClustering(linkage="ward", n_clusters=5)
# Run clustering
model.fit(ny_merged_scaled)
# Assign labels to main dataframe
ny_merge_2["ward5_label"] = model.labels_
# Assign labels to scaled dataframe
ny_merged_scaled_df["ward5_label"] = model.labels_
```

Next, let's look at the distribution of clusters when utilizing an AHC-based model. To do that you'll execute a groupby statement on the `"ward5_label"` variable, as seen in the next code block:

```
Ward5sizes = ny_merge_2.groupby("ward5_label").size()
Ward5sizes
```

Figure 8.11 displays the distribution of tracts by cluster:

Cluster Label	Number of Tracts
0	1,784
1	920
2	412
3	82
4	1,656

Figure 8.11 – AHC cluster tract count

You can then calculate the mean of the scaled data and plot the radial plot similar to what was done in the K-means exercise. We've opted to leave the code out for this task as it's largely similar to what was used to produce the last radial plot.

The resulting plot is displayed in *Figure 8.12*. From the plot, you can see that cluster **3** in this version is similar to cluster **1** in the K-means-based model, having a higher income and a more educated population.

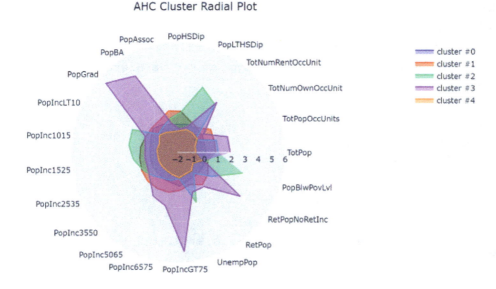

Figure 8.12 – AHC Cluster Radial Plot

Next, let's plot the spatial distribution of the AHC results next to the K-means-based results to see how they are similar and different across geographic space. To do that, you'll execute the next code block:

```
# Setup figure and ax
fig, axs = plt.subplots(2, 2, figsize=(12, 6))
# Plotting the k-means map
ax = axs[0,0]
# Plot the choropleth map of the k-means results
ny_merge_2.plot(
    column="kmeans_5_label",
    categorical=True,
    cmap="Set2",
    legend=True,
```

```
        legend_kwds={'loc': 'center left',
                      'bbox_to_anchor':(1,0.5),'fmt': "{:.0f}"},
        linewidth=0,
        ax=ax,
)
# Remove the axis
ax.set_axis_off()
# Add the title
ax.set_title("KMeans with $k=5$")
# Plot the choropleth map of the Agglomerative Hierarchical
Clustering results
ax = axs[0,1]
ny_merge_2.plot(
    column="ward5_label",
    categorical=True,
    cmap="Set3",
    legend=True,
    legend_kwds={'loc': 'center left',
                  'bbox_to_anchor':(1,0.5),'fmt': "{:.0f}"},
    linewidth=0,
    ax=ax,
)
# Remove the axis
ax.set_axis_off()
# Add the title
ax.set_title("AHC with $k=5$")
# Deleting the empty axis
axs[1,0].set_axis_off()
axs[1,1].set_axis_off()
# Display the map
plt.show()
```

Figure 8.13 shows the maps of the two clustering approaches. You'll notice that cluster **4** in the AHC model looks very similar to cluster **0** in the K-means model:

Figure 8.13 – AHC versus K-means clustering choropleth maps

Up until this point, we've walked you through two clustering methodologies that use spatial data in their development, but the formation of the clusters is not constrained by geography. In the next section, we'll introduce you to agglomerative clustering with a spatial constraint.

Spatially constrained agglomerative hierarchical geodemographic clustering

In this section, you'll be building out an AHC model that incorporates a spatial constraint. For some use cases, you may require your clusters to be more spatially connected, or constrained. Use cases could include setting service or management boundaries for an organization. This type of clustering is also known as **regionalization**, as it produces connected geographic clusters or regions.

To begin constructing the spatially constrained agglomerative hierarchical cluster model, you'll first need to decide on the spatial constraint, which is a spatial weights matrix. Prior to this chapter, we developed a spatial weights matrix to test for spatial autocorrelation. We'll leverage that weights matrix for this exercise.

The code in the next code block will look largely the same as the non-spatially constrained AHC model with the exception of the connectivity parameter. This parameter is now set to the spatial weights matrix, w, which was defined earlier in this chapter:

```
# Set the seed for reproducibility
np.random.seed(54321)
# Specify cluster model with a spatial constraint. Constraint
is passed using connectivity parameter.
model = AgglomerativeClustering(
    linkage="ward", connectivity=w.sparse, n_clusters=5
)
# Fit the algorithm to the data
model.fit(ny_merged_scaled)
# Assign the labels to dataframe
ny_merge_2["ward5wgt_label"] = model.labels_
```

```
# Assign labels to the scaled dataframe
ny_merged_scaled_df["ward5wgt_label"] = model.labels_
```

With the spatially constrained AHC model fit, you can now map it in comparison to the AHC and the K-means model. The code is largely similar to the last code block and for brevity has been left out of this section, but is included in the Jupyter notebook.

The resulting maps are displayed in *Figure 8.14*. From the maps, you can see that the spatially constrained model puts most of the census tracts into the green cluster, while the other five clusters are mostly concentrated in Long Island and NYC:

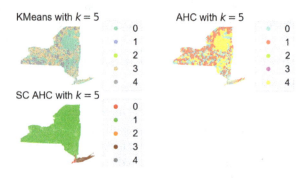

Figure 8.14 – SCAHC versus AHC versus K-means clustering choropleth maps

To conclude this section, we'll build one final model with a different spatial constraint. In this exercise, you'll set the spatial constraint to be the 10 **k-nearest neighbors** (**KNN**). To create this spatial constraint, you'll execute the next code block:

```
# Changing the spatial constraint to use KNN
w = KNN.from_dataframe(ny_merge_2, k=10)
```

You can now pass the KNN-based weights matrix to the agglomerative clustering module, which is done in the next code block:

```
# Setting the seed for reproducibility
np.random.seed(54321)
# Specifying the cluster model with KNN spatial constraint
model = AgglomerativeClustering(
    linkage="ward", connectivity=w.sparse, n_clusters=5
)
# Fitting the algorithm to the data
model.fit(ny_merged_scaled)
# Assigning the labels to dataframe
```

```
ny_merge_2["ward5_knnwgt_label"] = model.labels_
# Assigning labels to scaled dataframe
ny_merged_scaled_df["ward5_knnwgt_label"] = model.labels_
```

With the model built, it is now time to map the KNN-based model against the previous models. The resulting map is displayed in *Figure 8.15*. As you can see, the KNN-based approach yields fewer contiguous geographic regions than the SCAHC model using the queens-based weights matrix:

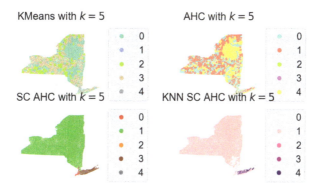

Figure 8.15 – KNN SCAHC versus SCAHC versus AHC versus K-means clustering choropleth maps

You've now calculated four different clustering options based on the geodemographic data for New York. Up until now, you've only assessed the clusters based on their profiling and visualizations. In the next section, you'll calculate various statistics to see how well the clusters perform mathematically.

Measuring model performance

There are a number of ways to measure the performance of a clustering model. Measuring the performance is important because it tells you how well each algorithm is doing in creating segments that have similar observations within each segment while appropriately differentiating segments from one another.

In this section, we'll review three common performance measures: the Calinski-Harabasz score, the Davies-Bouldin score, and the Silhouette score. Diving into each of these measures further is outside the scope of this book. To continue your learning, we recommend checking out the following resources mentioned in each of the points, as follows:

- **Calinski-Harabasz score**: Higher scores indicate better models (https://scikit-learn. org/stable/modules/generated/sklearn.metrics.calinski_harabasz_ score.html#sklearn.metrics.calinski_harabasz_score)

- **Davies-Bouldin score**: The minimum score is 0 and lower scores indicate better performance (https://scikit-learn.org/stable/modules/generated/sklearn.metrics.davies_bouldin_score.html)

- **Silhouette score**: Scores are bounded between -1 and 1 where a score of 1 indicates the perfect model performance (https://scikit-learn.org/stable/modules/generated/sklearn.metrics.silhouette_score.html#sklearn.metrics.silhouette_score)

To begin calculating the performance measures for the clustering models you have implemented, you'll need to import the methods for each metric. You'll then create three lists to store the calculated metric for each of the four models:

```
from sklearn.metrics import calinski_harabasz_score, davies_
bouldin_score, silhouette_score
ch_scores = []
db_scores = []
s_scores = []
```

Next, you'll loop over the models and calculate each of the metrics:

```
for model in ("kmeans_5_label", "ward5_label", "ward5wgt_
label", "ward5_knnwgt_label"):
    # compute the CH score
    ch_score = calinski_harabasz_score(
        ny_merged_scaled_df[geo_demo_rn],
        ny_merged_scaled_df[model],
    )
    ch_scores.append((model, ch_score))
    # compute the DB score
    db_score = davies_bouldin_score(
        ny_merged_scaled_df[geo_demo_rn],
        ny_merged_scaled_df[model],
    )
    db_scores.append((model, db_score))
    # compute the silhouette score
    s_score = silhouette_score(
        ny_merged_scaled_df[geo_demo_rn],
        ny_merged_scaled_df[model],
```

```
    )
    s_scores.append((model, s_score))
```

Lastly, you'll combine all of the calculated metrics into a single DataFrame for easy comparison:

```
# create a dataframe from the scores
ch_df = pd.DataFrame(
    ch_scores, columns=["model", "CH score"]
).set_index("model")
db_df = pd.DataFrame(
    db_scores, columns=["model", "DB score"]
).set_index("model")
s_df = pd.DataFrame(
    s_scores, columns=["model", "Silhouettescore"]
).set_index("model")
# Merging into a combined dataframe
scores_df = ch_df.merge(db_df, on="model")
scores_df = scores_df.merge(s_df, on="model")
# displaying the dataframe
scores_df
```

The resulting scores are displayed in *Figure 8.16*:

Model	CH Score	DB Score	Silhouette Score
kmeans_5_label	1346.75	1.39	0.24
ward5_label	1162.35	1.54	0.17
ward5wgt_label	281.13	3.20	0.01
ward5_knnwgt_label	446.28	2.08	0.12

Figure 8.16 – Measures of model performance

Based on the model scores, you can see that the non-spatially constrained models perform better mathematically than their spatially constrained counterparts across all measures. This is to be expected as the spatial constraint limits the ability of the clustering algorithms to minimize within-cluster variance while maximizing between-cluster variance. Depending on your use case, the ability to have geographically contiguous clusters may be more important than the mathematical performance of the model. Although, if mathematical performance is the top priority, you'd want to choose the K-means-based model as it has the best performance statistics.

Summary

In this chapter, we introduced you to a number of clustering algorithms, including K-means and AHC. We also introduced you to a variant of AHC that leverages a spatial constraint via the spatial weights matrix to develop geographically constrained clusters known as regions.

For each of the clustering models, we evaluated the clusters through cluster profiling. We produced maps of each cluster and also calculated a variety of descriptive statistics, including cluster tract counts, average cluster tract area, and mean values. We then used this information to produce choropleth maps of each of the clustering algorithms.

In the final section of the chapter, we introduced you to the Calinski-Harabasz score, the Davies-Bouldin score, and the Silhouette score, which are common mathematical measures of clustering performance. Even though the K-means-based model scored the best mathematically, it may not be the clustering model that makes the most sense for your use case. For some use cases, such as management structures or service areas, it may make sense to sacrifice mathematical performance for the added benefit of having geographically constrained clusters.

In *Chapter 9, Developing Spatial Regression Models*, we will introduce you to a number of regression algorithms that are built with geography in mind.

9
Developing Spatial Regression Models

In this chapter, we will be discussing regression models and how they can be improved by incorporating spatial structures. Spatial structures can be an important facet of building into traditional regression models, but they are often overlooked. It is important to consider spatial structures and to build them into a regression model when the process that generated the source data is geographic in nature.

To understand this better, it's helpful to think through a potential real-world situation. Imagine that you operate a chain of high-end furniture stores and you're trying to identify the best location for a future storefront that would maximize sales. Sales at your existing stores could be impacted by the number of cars that pass by the store every day, the proximity to other furniture stores, the number of new housing developments in nearby neighborhoods, and the affluence of the population in the vicinity. Each of these potentially explanatory correlating factors has a spatial component to them that could add value to the performance of a regression model.

In the preceding example, a potential regression model could be used in one of two ways:

- To **explain** the relationship between a variable and store sales. The explanation is important because it allows you to model a phenomenon or relationship to understand it better and leverage this understanding to make more informed decisions in the future.

- To **predict** future store sales. Prediction can be very powerful, as it allows you to infer the future value at a particular location or future time based on information from historical data.

In this chapter, we'll cover the following topics:

- A refresher on regression models
- How to incorporate space into regression models
- Identifying and working with spatial fixed effects

Technical requirements

For this chapter, you'll leverage the Jupyter notebook called `Chapter 9 - Spatial Regression`, which is located on the book's GitHub repo at `https://github.com/PacktPublishing/Applied-Geospatial-Data-Science-with-Python/tree/main/Chapter09`.

A refresher on regression models

It is best if we start with a brief refresher on regression models in general to ensure a common understanding. Let's begin with the following regression equation:

$$Y = \beta_0 + \beta_1 X_1 + \beta_2 X_2 + \cdots + \beta_n X_n + \varepsilon$$

Let's break down the notation in this equation:

- Y is the dependent variable, representing the process you are trying to explain or predict.

- β_0 is the intercept, which is the value of the dependent variable if all of the independent variables are 0.

- β_n, known as beta, represent the coefficients applied to the independent variables. These are computed by the regression algorithm and represent the strength and direction of the relationship between the independent and dependent variables.

- X_n are the independent or explanatory variables used to explain or predict the dependent variable.

- ε is the error term.

Now that we've aligned on a common understanding of the regression equation and terms, let's shift our focus to building out an initial regression model.

Constructing an initial regression model

You'll be constructing an **ordinary least squares** (**OLS**) regression model as the first model in this chapter. This model will be used as a baseline for comparison of other models constructed later on. For this chapter, we are going to continue leveraging the **New York City** (**NYC**) Airbnb dataset that we've worked with in prior chapters. Our focus for this analysis will be attempting to build an explanatory model for nightly Airbnb rental prices.

To build the OLS model, you'll execute the following steps:

1. Import the required packages:

```
from pysal.lib import weights
from pysal.explore import esda
import numpy as np
```

```
import pandas as pd
import geopandas as gpd
import matplotlib.pyplot as plt
import seaborn
import contextily
```

2. Read in the NYC Airbnb data:

```
# Reading in the data
data_path = r'YOUR FILE PATH'

listings = pd.read_csv(data_path + r'NY Airbnb June 2020\
listings.csv.gz', compression='gzip', low_memory=False)

#Converting it to a GeoPandas DataFrame:
listings_gpdf = gpd.GeoDataFrame(
    listings,
    geometry=gpd.points_from_xy(listings['longitude'],
                                listings['latitude'],
                      crs="EPSG:4326")
)
```

3. Now we'll focus on attractions in Manhattan, so we need to create a mask to filter locations in the Manhattan borough:

```
boroughs = gpd.read_file(data_path + r"NYC Boroughs\
nybb_22a\nybb.shp")
manhattan = boroughs[boroughs['BoroName']=='Manhattan']
manhattan = manhattan.to_crs('EPSG:4326')

listings_mask = listings_gpdf.within(manhattan.loc[3,
'geometry'])

listings_manhattan = listings_gpdf.loc[listings_mask]
```

4. With the data read in, you'll subset the data to only include a handful of variables that may be indicative of nightly rental rates for Airbnbs in NYC. Here, you'll select variables such as the neighborhood, the number of bedrooms and beds, and the room type:

```
#Subsetting the data to a handful of variables that could
be indicative of nightly Airbnb price
```

```
voi = ['id' # Unique identifier for the listing
       ,'room_type' # Type of room
       ,'accommodates' # The maximum capacity of the
listing
       ,'bedrooms' # The number of bedrooms
       ,'beds' # The number of beds
       ,'review_scores_rating' # The rating
       ,'price' # The nightly rental rate, dependent
variable (Y)
       ]


listings_manhattan_subset = listings_manhattan[voi]
```

5. As with all real-world examples, this dataset may need to be cleaned and preprocessed to be in the best shape for modeling. First, you'll want to check to see whether the data types are formatted in an intuitive way and in a way that the algorithm can understand:

```
#Checking the data types of the variables
listings_manhattan_subset.dtypes
```

The output from executing the previous code block is displayed in *Figure 9.1*.

```
id                        int64
room_type                object
accommodates              int64
bedrooms                float64
beds                    float64
review_scores_rating    float64
price                    object
dtype: object
```

Figure 9.1 – Data types

From the preceding output, you can quickly identify some potential issues which will need to be corrected. First, the room_type variable is stored as an object meaning that it is likely a categorical text object. You'll want to confirm this is the case and then **encode** the variable so that the algorithm can interpret it. Secondly, the price variable is also stored as an object. This doesn't make intuitive sense, as the price variable should be stored as a float type.

To explore the first issue in more detail, you'll run the next step in the analysis.

6. Identify the types of rooms stored in the `room_type` variable:

```
for col in listings_manhattan[['room_type']]:
    print(listings_manhattan_subset[col].unique())
```

Executing the code in the prior code snippet generates the following output:

```
['Entire home/apt' 'Private room' 'Shared room' 'Hotel
room']
```

There are four types of rooms in the dataset. In order for the algorithm to understand these room types, they'll need to be encoded as binary variables. We'll choose to encode the variables using **one-hot encoding**, as it is one of the most common and easiest ways to encode categorical variables.

7. One-hot encode the `room_type` variable:

```
# Encoding categorical variables
# Room Type, All 0s = hotel room
listings_manhattan_subset['rt_entire_home_apartment'] =
np.where(listings_manhattan_subset["room_type"]=='Entire
home/apt', 1, 0)
listings_manhattan_subset['rt_private_room'] =
np.where(listings_manhattan_subset["room_type"]=='Private
room', 1, 0)
listings_manhattan_subset['rt_shared_room'] =
np.where(listings_manhattan_subset["room_type"]=='Shared
room', 1, 0)
```

Next, we can explore the issue with the `price` variable.

8. Check the values stored in the `price` variable:

```
listings_manhattan_subset['price'].values
```

The output from the prior code snippet is:

```
array(['$225.00', '$68.00', '$75.00', ..., '$93.00',
'$462.00', '$113.00'], dtype=object)
```

After reviewing the output of the code snippet, you can see that the values are encased in quotes with a $ in front of each value. The `dtype` of the variable is an object type, indicating that this is a string and not a float. You'll need to remove the $ and convert the variable to a float. At the same time, you'll go ahead and log transform the `price` variable as well.

9. Correct the format of the `price` variable:

```
# Cleaning up the price column
listings_manhattan_subset['price'] = listings_manhattan_
subset['price'].str.replace('$', '')
listings_manhattan_subset['price'] = listings_manhattan_
subset['price'].str.replace(',', '')
listings_manhattan_subset['price'] = listings_manhattan_
subset['price'].astype(float)
# Logging the price variable
listings_manhattan_subset['log_price'] = np.log(listings_
manhattan_subset['price'])
```

10. Our last quality check will be to check for missing values in the data:

```
# Checking for missingness
print('Total Records:', len(listings_manhattan_subset))
listings_manhattan_subset.isna().sum()
```

The output of the prior code block is displayed in *Figure 9.2*.

```
Total Records: 15284
id                        0
room_type                 0
accommodates              0
bedrooms               2403
beds                    363
review_scores_rating   3738
price                     0
dtype: int64
```

Figure 9.2 – Missing values

From this output, you can see that roughly 15% to 25% of the records have a missing value for the `bedrooms` and `review_scores_rating` variables. A smaller percentage of the data is also missing a value for the `beds` variable. If it made intuitive sense to impute the missing value for these variables, then we would rather go that route so that we could retain as much information as possible. However, there is no logical imputation method, as each Airbnb likely has unique characteristics. As such, we'll move forward by dropping the missing records.

11. Drop records with missing data:

```
listings_manhattan_subset = listings_manhattan_
subset[listings_manhattan_subset['bedrooms'].notna()]
listings_manhattan_subset = listings_manhattan_
subset[listings_manhattan_subset['review_scores_rating'].
notna()]
listings_manhattan_subset = listings_manhattan_
subset[listings_manhattan_subset['beds'].notna()]
listings_manhattan_subset.isna().sum()
```

Now that you have a clean dataset, you can begin building out an OLS regression model. To begin, you'll need to import an additional library called `spreg` from PySAL's `model` package. **Spreg** stands for **spatial regression** and is a Python package that allows you to estimate models that include spatial components.

12. Importing the `spreg` package:

```
from pysal.model import spreg
```

Next, you'll want to define a list of variables that will be used to explain the nightly Airbnb rental rate.

13. Defining explanatory variables:

```
# Defining a list of explanatory variables
m_vars = ['accommodates', 'bedrooms', 'beds',
          'review_scores_rating',
          'rt_entire_home_apartment',
          'rt_private_room', 'rt_shared_room'
         ]
```

You can now pass the list of explanatory variables to the `spreg.OLS` function to build your initial model and display the results.

14. Constructing the OLS model:

```
ols_m = spreg.OLS(
    listings_manhattan_subset[["log_price"]].values # the
dependent variable (Y)
    ,listings_manhattan_subset[m_vars].values # the
independent variables(Xs)
    ,name_y = 'price',
```

```
        name_x = m_vars
    )
    print(ols_m.summary)
```

A snippet of the OLS output is displayed in *Figure 9.3*.

```
REGRESSION
----------
SUMMARY OF OUTPUT: ORDINARY LEAST SQUARES
-----------------------------------------
Data set            :    unknown
Weights matrix      :       None
Dependent Variable  :      price      Number of Observations:       9471
Mean dependent var  :     5.1119      Number of Variables   :          8
S.D. dependent var  :     0.7428      Degrees of Freedom    :       9463
R-squared           :     0.4096
Adjusted R-squared  :     0.4092
Sum squared residual:   3084.716      F-statistic           :   937.8547
Sigma-square        :      0.326      Prob(F-statistic)     :          0
S.E. of regression  :      0.571      Log likelihood        :  -8126.601
Sigma-square ML     :      0.326      Akaike info criterion :  16269.203
S.E of regression ML:     0.5707      Schwarz criterion     :  16326.451


---------------------------------------------------------------------------
          Variable    Coefficient      Std.Error    t-Statistic   Probability
---------------------------------------------------------------------------
          CONSTANT      5.0991240      0.0645378     79.0099165     0.0000000
      accommodates      0.1474902      0.0057503     25.6490938     0.0000000
          bedrooms      0.1021910      0.0138345      7.3866977     0.0000000
              beds      0.0031612      0.0098316      0.3215343     0.7478126
review_scores_rating    0.0563580      0.0071805      7.8487573     0.0000000
rt_entire_home_apartment  -0.6283971   0.0551371    -11.3970011     0.0000000
    rt_private_room     -1.0887888      0.0553499    -19.6710123     0.0000000
     rt_shared_room     -1.2999558      0.0709383    -18.3251493     0.0000000
---------------------------------------------------------------------------


REGRESSION DIAGNOSTICS
MULTICOLLINEARITY CONDITION NUMBER          32.655

TEST ON NORMALITY OF ERRORS
TEST                      DF        VALUE          PROB
Jarque-Bera                2    10206.015        0.0000

DIAGNOSTICS FOR HETEROSKEDASTICITY
RANDOM COEFFICIENTS
TEST                      DF        VALUE          PROB
Breusch-Pagan test         7      185.221        0.0000
Koenker-Bassett test       7       56.264        0.0000
=============================== END OF REPORT =====================================
```

Figure 9.3 – Initial model OLS output

While a complete explanation of the OLS report is beyond the scope of this text, we will focus on a few important sections. First is the **Adjusted R-squared** value of 0.4092. The Adjusted R-squared value is a measure of the goodness of fit of the model in terms of its ability to explain the variance in the dependent variable. In contrast to the **R-squared** value, the Adjusted R-squared value is negatively impacted as the number of independent variables in the model increase if the additional variable does not benefit the model. The range of R-squared and Adjusted R-squared is between 0 and 1, where a value closer to 1 indicates better goodness of fit. Our initial model does not explain a large percentage of the underlying variance, and this is something we'll work to improve throughout the chapter. To learn more about the OLS output, you can review PySAL's spreg documentation by visiting `https://pysal.org/spreg/generated/spreg.OLS.html`.

In the next section, we'll focus on the coefficients. As a reminder, the coefficients indicate the strength and direction of the relationship between the independent and dependent variables holding all else constant. The coefficients displayed in *Figure 9.3* generally align with expectations. Airbnbs that accommodate more people, have more bedrooms, and have more beds are more expensive than those that can accommodate fewer people, have fewer bedrooms, and have fewer beds. Airbnbs that have higher reviews are also more expensive than those with lower reviews. Finally, the categorical variables indicate that hotels are more expensive than other Airbnb types, with rooms that are shared or a private room in a shared building being less expensive.

With the initial regression model built, you can now begin exploring it a bit deeper to see whether there are any unmodeled spatial relationships.

Exploring unmodeled spatial relationships

In the initial model, we purposely left out spatial attributes, but this may have been done to the detriment of the model's ability to accurately explain the average nightly Airbnb price. In this section, you'll begin diving deeper into the model to check for any hidden spatial relationships.

One piece of information that is in the initial dataset is a variable called `neighbourhood_cleansed`, which indicates the neighborhood where the Airbnb is located. Some people may be willing to pay more to be located in a certain neighborhood, as it may be closer to an attraction they want to visit or a restaurant they wish to dine at. To explore this further, you'll calculate the average value of the residual by neighborhood and then produce a violin plot to display the residual distributions:

1. Explore model residuals by neighborhood:

```
# Creating a column to store the model residuals
listings_manhattan_subset["ols_m_r"] = ols_m.u
```

```
# Bringing back in the neighborhood variable
listings_manhattan_subset = listings_manhattan_subset.
merge(listings_manhattan[['id','neighbourhood_
cleansed']], how='left',on='id')
```

2. Calculate the average value of the residual by neighborhood:

```
mean = (
    listings_manhattan_subset.groupby("neighbourhood_
cleansed")
    .ols_m_r.mean()
    .to_frame("neighborhood_residual")
)
```

3. Create a DataFrame to store the residuals by neighborhood:

```
residuals_neighborhood = listings_manhattan_subset.merge(
        mean, how="left", left_on="neighbourhood_
cleansed", right_index=True
    ).sort_values("neighborhood_residual")
import plotly.express as px
import plotly.io as pio
from IPython.display import HTML
```

4. Plot the distribution of the residuals in a violin plot:

```
fig = px.violin(residuals_neighborhood,
                x="neighbourhood_cleansed",
                y="ols_m_r",
                color="neighbourhood_cleansed" )
# Updating the x and y axis labels
fig.update_layout(
    xaxis_title="Neighborhood",yaxis_title="Residuals"
)
HTML(fig.to_html())
```

The resulting violin plot is displayed in *Figure 9.4*.

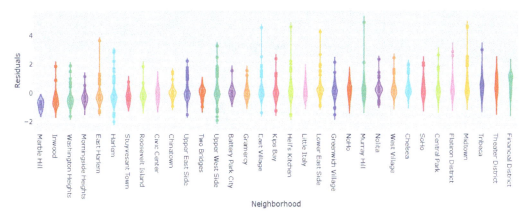

Figure 9.4 – Violin plot of neighborhood residuals

From the violin plot, you can see that some neighborhoods have higher residuals than others, including Midtown, the Theater District, and the Financial District, which are generally more desirable neighborhoods for tourists. This indicates that there may be an underlying spatial structure that we need to account for within the model.

Another thing that you'll want to check for is how the residuals distribute across space, as the residuals may cluster together in some fashion that you can't pick up on from the preceding violin plot. You'll run the next step to produce a map of the average residual by neighborhood.

5. Produce a neighborhood map of model residuals:

```
# Reading in the NYC Neighborhoods shapefile from data
file
import os
# Path
path = r"YOUR FILE PATH"


nyc_n = gpd.read_file(os.path.join(path,r"NYC
Neighborhoods\NYC_Neighborhoods.geojson"),
driver='GeoJSON')
```

6. Merge the datasets together:

```
nyc_n_r = nyc_n.merge(mean, left_on='neighborhood',
right_on="neighbourhood_cleansed")
```

7. Plot a choropleth of the model residuals:

```
f, ax = plt.subplots(1, figsize=(10, 10))
nyc_n_r.plot(
    column='neighborhood_residual',
    cmap='vlag',
    scheme='quantiles',
    k=4,
    edgecolor='white',
    linewidth=0.1,
    alpha=0.75,
    legend=True,
    ax=ax
)
```

8. Add a contextily basemap to add context to the map:

```
contextily.add_basemap(
    ax,
    crs=nyc_n.crs,
    source=contextily.providers.Stamen.Watercolor,
)

for idx, row in nyc_n_r.iterrows():
    plt.annotate(text=row['neighborhood'],
                 xy=tuple([row.geometry.centroid.x,
                           row.geometry.centroid.y]),
                 horizontalalignment='center',
                 fontsize=8, rotation=90)
# Remove axis
ax.set_axis_off()
# Displaying the map
plt.show()
```

Figure 9.5 shows the average residual by neighborhood.

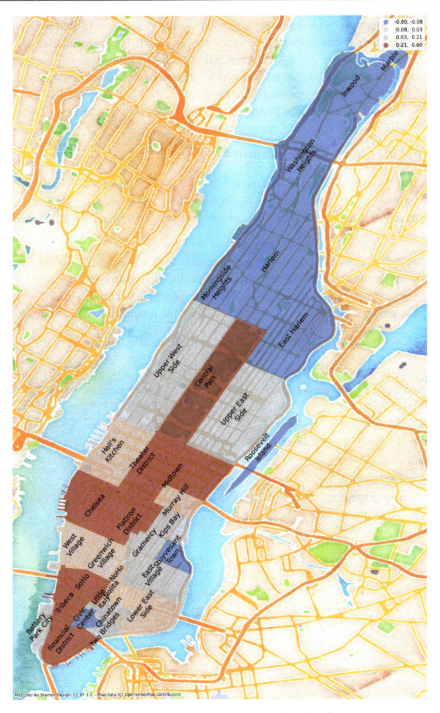

Figure 9.5 – Model residuals by neighborhood

Based on this figure, you can see that the residuals in one neighborhood tend to be similar to that of proximate neighborhoods. For instance, the residuals in the Financial District, Tribeca, Soho, and Little Italy are all on the higher end, with the neighborhoods being next to one another. This grouping of higher residuals could represent some form of geographic creep in the residuals from one neighborhood to another. This is likely due to Airbnb operators pricing their Airbnbs similar to Airbnbs in nearby neighborhoods. At this point, our model doesn't have the ability to pick up on this geographic signal.

Using spatial lags to check for geographic structure

You can look into the potential geographic structure in more detail by creating a **k-nearest neighbors (KNN)**-based spatial weights matrix, which will be used to construct a spatial lag. As a refresher, you can revisit spatial weights matrices, which we covered in *Chapter 6, Hypothesis Testing and Spatial Randomness*, before moving on to the next step.

In the next step, you'll construct a spatial weights matrix to look at the residuals of the five nearest Airbnbs and compare the spatial lag to the model residual for each Airbnb observation. Looking at the residuals helps us understand if there is a spatial pattern present in the model's error. On the *y* axis, you'll plot the lagged residuals, and on the *x* axis, you'll plot the residual for an individual location:

1. Build a KNN spatial weights matrix based on the five nearest neighbors:

    ```
    # Bringing in the geometry attribute
    residuals_neighborhood = residuals_neighborhood.
    merge(listings_manhattan[['id','geometry']],
    how='left',on='id')
    # Building the spatial weights matrix
    knn = weights.KNN.from_dataframe(residuals_
    neighborhood,k=5)
    # Constructing the spatial lag
    lag_residual = weights.spatial_lag.lag_spatial(knn,
    ols_m.u)
    ```

2. Plot the results of the spatial lag:

    ```
    fig = px.scatter(x=ols_m.u.flatten(),
                     y=lag_residual.flatten(),
                     trendline="ols",
                     width=800, height=800)
    # Updating the x and y axis labels
    fig.update_layout(
        xaxis_title="Airbnb Residuals",
    ```

```
        yaxis_title="Spatially Lagged Residuals"
)
fig.show()
```

The resulting plot is shown as follows:

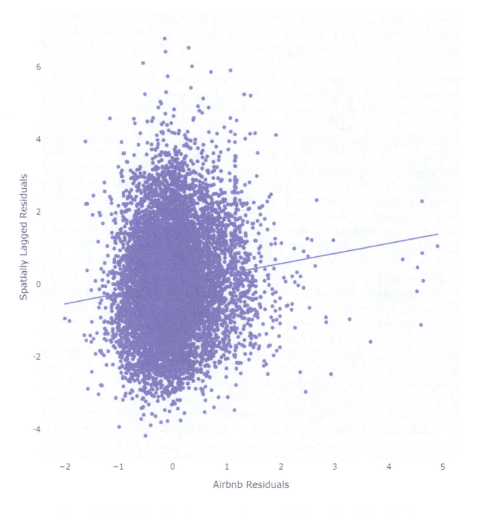

Figure 9.6 – Model residuals compared to spatially lagged model residuals

From this plot, you can see that the model residuals all tend to cluster tightly together, meaning that if an Airbnb's nightly price is over or under-predicted, then those around it are also likely to be over or under-predicted in a similar fashion. Now that you've explored some of the unmodeled spatial relationships in the data, it is time to start teaching the model how to think spatially with the inclusion of spatial features.

Teaching the model to think spatially

We kicked this chapter off with a brief disclaimer that it is important to consider spatial structures and incorporate them into the regression modeling process. This is especially important if the underlying data is generated via a geospatial process. Thankfully, there are numerous methods by which you can accomplish this. In this section, we will build spatial structures into our models in two ways. First, we'll incorporate some of the spatially engineered variables that were constructed in *Chapter 7, Spatial Feature Engineering*. The second way we will build space into the model is by exploring **spatial fixed effects**, and we'll talk more about this later on.

To begin, let's go ahead and bring the spatially engineered variables into the equation. In the following first step, you'll rerun the feature engineering process previously conducted to bring in the distance to some common NYC attractions:

1. Recreate spatially engineered variables:

    ```
    # Reading in data on popular NYC Attractions
    path2 = r'YOUR FILE PATH'
    nyc_attr = pd.read_csv(path2 + 'NYC Attractions.csv')
    # Convert PDF to GPDF
    nyc_attr_gpdf =  gpd.GeoDataFrame(
        nyc_attr,
        geometry=gpd.points_from_xy(nyc_attr['Longitude'],
                                    nyc_attr['Latitude'],
                                    crs="EPSG:4326")
    )
    ```

2. Calculate the distance to each attraction from each Airbnb:

    ```
    attractions = nyc_attr_gpdf.Attraction.unique()
    # Converting to a projected coordinate system
    nyc_attr_gpdf_p = nyc_attr_gpdf.to_crs('EPSG:2263')
    listings_manhattan_p = listings_manhattan.to_
    crs('EPSG:2263')
    # Applying a lambda function that calls geopandas
    distance function to calculate the distance between each
    Airbnb and each attraction
    distances = listings_manhattan_p.geometry.apply(lambda g:
    nyc_attr_gpdf_p.distance(g))
    ```

3. Rename the columns based on the attraction for which the distance is calculated:

    ```
    distances.columns = attractions
    ```

4. Convert the distance units from the US survey foot to miles:

```
distances = distances.apply(lambda x: x/5280, axis=1)
```

5. Create new variables to store the number of locations that are less than n miles from the Airbnb:

```
distances_1mi = distances.apply(lambda x: x <=1, axis=1).
sum(axis=1)
distances_2mi = distances.apply(lambda x: x <=2, axis=1).
sum(axis=1)
distances_3mi = distances.apply(lambda x: x <=3, axis=1).
sum(axis=1)
distances_4mi = distances.apply(lambda x: x <=4, axis=1).
sum(axis=1)
distances_5mi = distances.apply(lambda x: x <=5, axis=1).
sum(axis=1)
distances_6mi = distances.apply(lambda x: x <=6, axis=1).
sum(axis=1)
```

6. Create a DataFrame combining all the distance bands:

```
distance_df = pd.concat([distances_1mi,distances_2mi,
                         distances_3mi,distances_4mi,
                         distances_5mi,distances_6mi],
                         axis=1)
distance_df.columns =
['Attr_1mi','Attr_2mi','Attr_3mi','Attr_4mi',
'Attr_5mi','Attr_6mi']


# Joining back to the listings geopandas df
listings_manhattan = listings_manhattan.merge(distances,
left_index=True, right_index=True)
listings_manhattan = listings_manhattan.merge(distance_
df, left_index=True, right_index=True)
```

7. Next, add the spatially engineered variables to the list of explanatory variables defined earlier:

```
# Newly Created Geographic Features
g_vars = ['Central Park','Central Park Zoo',
          'Empire State Building','Statue of Liberty',
          'Rockeffeller Center','Chrysler Building',
          'Times Square', 'MoMa', 'Charging Bull']
```

```
# Bring the geographic features into our subset
listings_manhattan_subset = listings_manhattan_subset.
merge(listings_manhattan[['id','Central Park', 'Central
Park Zoo', 'Empire State Building', 'Statue of Liberty',
            'Rockeffeller Center', 'Chrysler Building',
'Times Square', 'MoMa', 'Charging Bull']],
how='left',on='id')


# Adding geographic features to model variables used
previously
g_m_vars = m_vars + g_vars
```

With the new list of explanatory variables declared, you can now build a second OLS model incorporating the newly created spatial features.

8. Build an OLS model with spatial features:

```
ols_m_g = spreg.OLS(
    listings_manhattan_subset[["log_price"]].values,
    listings_manhattan_subset[g_m_vars].values,
    name_y = 'price',
    name_x = g_m_vars
)
print(ols_m_g.summary)
```

Figure 9.7 shows a snippet of the first half of the summary report for this model describing the model performance statistics.

```
REGRESSION
----------
SUMMARY OF OUTPUT: ORDINARY LEAST SQUARES
-----------------------------------------
Data set            :     unknown
Weights matrix      :        None
Dependent Variable  :       price    Number of Observations:       9471
Mean dependent var  :      5.1119    Number of Variables   :         17
S.D. dependent var  :      0.7428    Degrees of Freedom    :       9454
R-squared           :      0.5439
Adjusted R-squared  :      0.5431
Sum squared residual:    2383.177    F-statistic           :   704.5267
Sigma-square        :       0.252    Prob(F-statistic)     :          0
S.E. of regression  :       0.502    Log likelihood        :  -6904.724
Sigma-square ML     :       0.252    Akaike info criterion :  13843.449
S.E of regression ML:      0.5016    Schwarz criterion     :  13965.101
```

Figure 9.7 – Spatial model performance statistics

The OLS output shows that the Adjusted R-squared statistic for this model is ~0.54, compared to a previous Adjusted R-squared statistic of ~0.40. Adding in the spatially engineered variables improved the model's ability to explain the variance in nightly Airbnb prices by 35%, which is a huge improvement.

9. You'll now want to check to see whether the inclusion of these features addressed all of the spatial structures present in the data. To do this, we'll construct a second residual plot similar to what was done in the last section. The code to do this is included in the Jupyter notebook but left out of this section for brevity and because it is largely similar to the last residual plot.

Figure 9.8 shows the resulting residual plot.

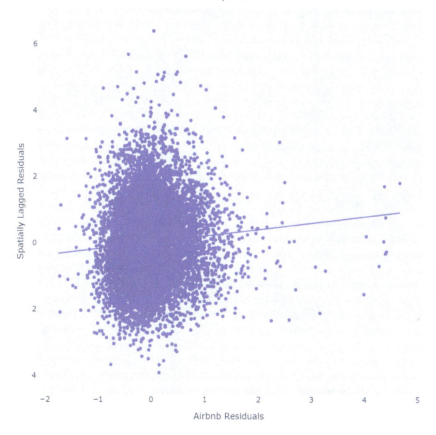

Figure 9.8 – Residual plot for the OLS model with geospatial features

We can see that there is still a spatial structure in our data as the residuals are still clustered together. We'll explore how to address the additional spatial structure through the incorporation of spatial fixed effects.

Incorporating spatial fixed effects within the model

The incorporation of spatial proximity features was your first foray into teaching the model to think spatially. While these features benefited model performance and addressed the underlying premium visitors may be willing to pay to stay in close proximity to famous NYC attractions, they did not address all of the spatial structure present in the data.

It's now time to think a bit more broadly about what may be behind the remaining spatial structure present in the data. As with neighborhoods across the country, underlying real estate values in NYC may vary from neighborhood to neighborhood. This underlying price structure in neighborhood real estate values could cause the host to pass along some or all of that cost to the potential renter of their Airbnb. This variation across space between the outcome variables and the explanatory, or predictor, variables is known as **spatial nonstationarity**. There are a few ways that we can account for this nonstationarity to hopefully improve our model, which we'll cover in the next few sections.

One way to account for spatial nonstationarity is by adding spatial fixed effects. A **fixed effect** is a term commonly found in econometrics and is defined as an effect that is held constant within some cross-section of the data while other model parameters are allowed to vary. In the next coding exercise, you'll build a spatial fixed effects model by calling the `OLS_Regimes` function from `spreg`. You'll pass the `neighbourhood_cleansed` variable to the function, which indicates the neighborhood each Airbnb falls within:

1. Perform a join to bring in the neighborhood variable:

    ```
    # Bringing back in the neighbourhood_cleansed variable
    listings_manhattan_subset = listings_manhattan_subset.
    merge(listings_manhattan[['id','neighbourhood_
    cleansed']], how='left',on='id')
    ```

2. Implement the spatial fixed effects model:

    ```
    sfe_m = spreg.OLS_Regimes(
        # The dependent variable (Y) - Log Price (log_price)
        listings_manhattan_subset[["log_price"]].values,

        # The independent variables (Xs)
        listings_manhattan_subset[g_m_vars].values,
        # Variable which specifies which neighborhood
    each airbnb falls within    listings_manhattan_
    subset["neighbourhood_cleansed"].tolist(),
        # Vary constant by each cross-section/group
        constant_regi="many",
        # Here we tell the model that the variables are kept
    ```

```
constant by group
    cols2regi=[False] * len(g_m_vars),
    # Here we tell the model to estimate a single sigma
coefficient
    regime_err_sep=False,
    # Dependent variable name
    name_y="log_price",
    # Independent variables names
    name_x=g_m_vars,
)
# Printing the model summary
print(sfe_m.summary)
```

Figure 9.9 shows the performance statistics for this model from the summary output.

```
REGRESSION
----------
SUMMARY OF OUTPUT: ORDINARY LEAST SQUARES - REGIMES
--------------------------------------------------
Data set            :     unknown
Weights matrix      :        None
Dependent Variable  :   log_price      Number of Observations:      9471
Mean dependent var  :      5.1119      Number of Variables   :        49
S.D. dependent var  :      0.7428      Degrees of Freedom    :      9422
R-squared           :      0.5632
Adjusted R-squared  :      0.5610
Sum squared residual:    2281.917      F-statistic           :  253.1438
Sigma-square        :       0.242      Prob(F-statistic)     :         0
S.E. of regression  :       0.492      Log likelihood        : -6699.114
Sigma-square ML     :       0.241      Akaike info criterion : 13496.228
S.E of regression ML:      0.4909      Schwarz criterion     : 13846.872
```

Figure 9.9 – Spatial fixed effects model performance measures

We can see that the inclusion of spatial fixed effects within the model once again improved the Adjusted R-squared, which now sits at ~0.56, compared to ~0.54 in the prior model.

In addition to the summary statistics on model performance and the coefficients of the model, there is a new section at the bottom called Regimes Diagnostics – Chow Test. The Chow test is performed to tell us whether the inclusion of the groups or regimes is statistically significant and produces different results. The value for the Chow test for this model is 418.104 with an incredibly low p-value of ~8.49e-69, indicating statistically different results by neighborhood.

Next, plot the residuals from the spatial fixed effects model. The residuals from the spatial fixed effects model are displayed in *Figure 9.10*.

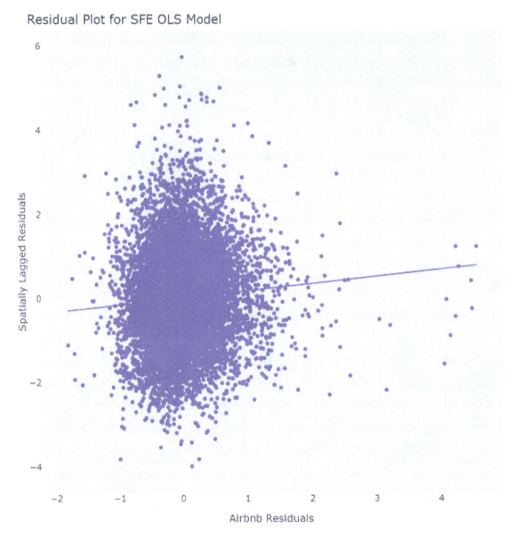

Figure 9.10 – Residual plot for spatial fixed effects model

Fixed effects models are one possible way to address spatial nonstationarity. Here, adding the fixed effects helped improve the residuals plot for this model with the lagged residuals now being bounded by -4 and 6 compared to prior models which exceeded 6. The Adjusted R-squared statistic for this model is ~0.56, compared to ~0.54 in the previous model, which indicates that you can now account for 56% of the variation in nightly Airbnb prices.

There is another class of models known as **Geographically Weighted Regression (GWR)**, which helps address spatial nonstationarity further by fitting local regression models for each independent, explanatory, variable. We'll discuss this in more detail in the next section and hope that we see a better model performance.

Introduction to GWR models

GWR models vary from OLS-based models in that instead of fitting a set of global estimates, GWR examines the way in which the relationship between each predictor variable varies across space with respect to the dependent variable. GWR does this by iteratively fitting a localized regression within a search window or neighborhood around each observation. The observation for which the regression is being fit is known as the **regression point**. Observations that are closer to the regression point are weighted more heavily in the regression than observations that are further away.

Fitting a regression within these local neighborhoods is performed by using either a **fixed kernel** or an **adaptive kernel**. A fixed kernel uses an identical search area across all regression points, while an adaptive kernel's search area can vary across space. *Figure 9.11* shows a fixed kernel approach compared to an adaptive kernel approach.

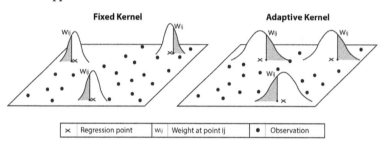

Figure 9.11 – Fixed and adaptive kernels

The search area is based on the **bandwidth** parameter, which is essentially an upward bound, after which the weight assigned to the observation points is set to zero. Setting a larger bandwidth means that more observations will be included in the fitting of the local regression model. One critical assumption made with GWR models is that it assumes that the scale of the neighborhood is identical for each of the independent variables. As such, the optimal bandwidth determined by the algorithm will be the same for all independent variables.

Let's transition now to fitting a GWR model using the information you just learned about.

Fitting a GWR model to predict nightly Airbnb prices

To begin fitting a GWR model, you need to import a few more modules:

1. Import the modules used for GWR models:

    ```
    from mgwr.gwr import GWR
    from mgwr.sel_bw import Sel_BW
    ```

    Next, you'll need to set up the inputs to the model, which are an array of the explanatory variables, an array of the dependent variable, and lastly, an array of the coordinates for each observation.

2. Set up the inputs for the model:

```
# Setting the explanatory variables
exp_vars = listings_manhattan[['accommodates',
'bedrooms','beds','review_scores_rating']].values
# Setting the dependent variable: log_price
y = (listings_manhattan['log_price'].values).reshape((-
1,1))
# Defining the coordinates of the observations
coords = list(zip(listings_manhattan.geometry.x,listings_
manhattan.geometry.y))
```

With the data formatted appropriately, it is now time to determine the optimal bandwidth for the kernel. To do that, you'll leverage the Sel_BW module. You'll pass the three arrays defined in the previous code snippet to the module.

3. Select the optimal bandwidth for the explanatory variables:

```
gwr_selector = Sel_BW(coords, y, exp_vars,
spherical=True)
gwr_bw = gwr_selector.search(bw_min=2)
# Displaying the optimal bandwidth
gwr_bw
```

Running the previous code snippet reveals that the optimal bandwidth for the explanatory variables is 156.

The next step is to fit the GWR model by passing the three data arrays and the optimal bandwidth that was determined in the last code snippet to the GWR function.

4. Fit the GWR model to your data:

```
gwr_results = GWR(coords, listings_manhattan[["log_
price"]].values,
                    exp_vars, gwr_bw).fit()
# Print the results of the GWR model
print(gwr_results)
```

5. You can then view the model results by calling the .summary() method:

```
gwr_results.summary()
```

The results for the GWR model, produced by executing the prior code snippet, are included in the Jupyter notebook and left out of this section for brevity. The model has an Adjusted R-squared value of 0.586, which is a slight improvement over the OLS model with fixed effects.

We'll not transition our discussion to a more advanced and newly developed version of GWR called Multiscale Geographically Weighted Regression.

Introduction to Multiscale Geographically Weighted Regression

Multiscale Geographically Weighted Regression (MGWR) is a more recent development in spatial regression modeling. In contrast to GWR, MGWR relaxes the assumption that the neighboring scale for each explanatory variable is identical and allows the scale of the analysis to vary for each of the explanatory variables. As such, an optimal bandwidth parameter is selected for each of the independent variables within an MGWR model.

MGWR is great for working with large datasets where the patterns between the independent and dependent variables vary across space and scale. However, this increased sophistication in the technique leads to increased computational complexity and, therefore, drastically increased runtime in fitting the model.

Fitting an MGWR model to predict nightly Airbnb prices

The process of fitting an MGWR model is very similar to that of GWR with a few modifications. For starters, you'll now leverage the MGWR module instead of the GWR model. Let's begin by importing that module into the notebook:

1. Import the MGWR module:

```
from mgwr.gwr import MGWR
```

Similar to GWR, you'll use the `Sel_BW` function to select the optimal bandwidth. However, you'll now set the `multi` parameter to `True`, indicating that this is a MGWR with different bandwidths for the explanatory variables.

2. Find the optimal bandwidths per explanatory variable:

```
selector = Sel_BW(coords, y, exp_vars, multi=True,
spherical=True)
selector.search(multi_bw_min=[4])
```

3. Lastly, run the MGWR function to fit the model and display the results:

```
mgwr_results = MGWR(coords, y, exp_vars, selector,
sigma2_v1=True).fit()
# Print the results of the GWR model
print(mgwr_results)
mgwr_results.summary()
```

The MGWR model is a step back in terms of model performance, with an Adjusted R-squared of only 0.384. This is likely because our data does not have variation across scales or require varying bandwidth for each of the covariates.

One additional thing to note about MGWR models as they're currently implemented is that they do not handle binary explanatory variables. If you pass a binary variable to an MGWR model, the model can exhibit the following behavior:

- It can take an extraordinary amount of time to find the optimal bandwidth

- The model results will be uninterpretable

In the Jupyter notebook, there is a section titled *MGWR with Binary Variables*. This model takes roughly 32 hours to run. We've included this section so that you have an example of what happens when binary explanatory variables are input into the model and the outcomes.

In the next section, we'll briefly touch on how you should think about selecting between these various types of models while balancing computational intensity with model performance.

How do I choose between these models?

When choosing between these various types of models, it is important to understand your data as well as the assumptions that go into each of the various models. It is also important to balance model performance with your individual operational constraints, such as how long you're willing to wait for model results. Here are a few questions that you can ask yourself to help determine which model may be better for each situation:

- Do the patterns between my target and explanatory variables vary across space?

  - If the answer to this question is yes, then fitting a GWR or an MGWR model may be a better-suited option, as OLS fits a global regression compared to the local regression fit by GWR and MGWR.

- If the patterns between my target and explanatory variables vary across space, do they also operate at different scales?

  - If the answer to these questions is yes, then MGWR is a better-suited option than GWR. Recall that GWR assumes a single scale for all explanatory variables, while MGWR relaxes this assumption.

- How long am I willing to wait for the model to run?

  - As your dataset grows in size, the time it takes to fit a GWR and MGWR model will increase. You'll need to balance model performance with operational constraints related to the time it takes to fit the model and deliver your results.

Let's transition now to summarize what you've learned throughout this chapter.

Summary

In this chapter, we walked you through the construction of spatial regression models to better understand the drivers of nightly Airbnb prices in NYC. We started the chapter off with a refresher on OLS regression models. Using this model, we looked at the distribution of the model's residuals to better understand some latent spatial structures that needed to be accounted for.

In the second section, you learned how to incorporate spatially engineered proximity features into the model, which dramatically improved the model's performance. We then introduced you to spatial fixed effects and how to use the spreg library's `OLS_Regimes` function to build a spatial fixed effects model, which further improved performance. Within this section, we also introduced the Chow test to ensure that the neighborhoods yielded statistically different results.

In the second section, you learned about GWR and MGWR, which are models that fit local regressions for each observation. MGWR builds upon GWR by relaxing the assumption that each explanatory model operates within the same scale.

We concluded the chapter with a brief discussion on when you should apply the various types of models by taking into account the model's performance and the need to consider whether the explanatory variables and their relationship with the target variable vary across space. We also discussed the operational complexity of fitting an MGWR model, which can run for several hours to several days.

In *Chapter 10, Developing Solutions for Spatial Optimization Problems*, we'll show you how to identify optimal solutions to a number of problems, including vehicle routing and location selection. Each of these problems requires spatial inputs to find the best option while operating around a number of constraints. We hope you're looking forward to learning how spatial data science can help you problem solve and is a critical component of how the real world operates.

10
Developing Solutions for Spatial Optimization Problems

Have you ever thought about what goes into ensuring that the packages you order online make it to your doorstep within their specified delivery window? Do you ever worry about the ability of your community's fire, police, or emergency medical services to arrive promptly when you or your family is in need? The last time you took a flight, did you think about the complex logistics happening behind the scenes to ensure that your pilot, flight attendants, and even the plane made it to the right place at the right time so that you could get to your destination? Each of these scenarios presents us with an optimization problem that is founded on geography. In fact, in our modern society, a growing number of problems require spatially optimized solutions.

In this chapter, we will cover the topic of spatial optimization, which is a set of incredibly powerful spatial analysis tools that are used to find optimal or near-optimal solutions. These solutions are based on minimizing or maximizing an objective function while operating within a set of constraints. Many types of spatial optimization problems can be solved through several methods, from exact techniques such as **linear programming** to heuristic-based approaches such as **greedy algorithms** and **genetic algorithms**.

Spatial optimization problems, similar to more general optimization problems, fall within a class of problems called **NP-hard**. This means, they are nondeterministic polynomial time-hard problems. Problems with relatively simple objective functions and very few constraints can yield a complex potential solution space and are thus computationally intensive to solve. Furthermore, the objective function and constraints that model this problem space can often be quite difficult to program.

This chapter could be an entire book on its own, given the number of problem types and the number of potential techniques to solve them. For brevity, we'll cover three types of problems: the **Location Set Covering Problem** (**LSCP**), the **Traveling Salesperson Problem** (**TSP**), and the **Capacitated Vehicle Routing Problem** (**CVRP**) – these are some of the most common spatial optimization problems that you are likely to encounter. The example mentioned previously regarding package delivery is a CVRP-based problem, while the emergency service example is an example of an LSCP-based problem.

In this chapter, you will learn about the following:

- How to identify spatial optimization problems
- How to program objective functions and constraints to identify optimal solutions
- How to interface with the Google Maps API for Python

Technical requirements

In this chapter, you'll leverage the `Chapter 10 - Spatial Optimization.ipynb` Jupyter notebook, which is stored in this book's GitHub repository at `https://github.com/ PacktPublishing/Applied-Geospatial-Data-Science-with-Python/tree/ main/Chapter10`. You'll also need your Google Maps API key, which was discussed in the preface of this book.

Exploring the Location Set Covering Problem (LSCP)

LSCPs fall within a class of problems known as **set covering problems**. The set covering problem class is an NP-hard problem within the **combinatorial optimization** space. These problems typically aim at minimizing the number of sets within a space that cover all the demand within that space. Take, for example, the emergency services problem we mentioned at the start of this chapter, where a given number of emergency service facilities must service the demand of all the residents in a given community. The emergency planning departments within this community may want to know if the demand from the community can be met with the existing infrastructure. If there are too few facilities to triage the community's needs, then additional facilities may need to be constructed. Conversely, there may be too many facilities in the community than needed. This may present opportunities to consolidate the facilities, resulting in operational savings.

In the case of emergency services, simply serving the demand is not enough. These services are often responding to emergencies that are time sensitive, such as when someone is having a heart attack or when a house is on fire. A couple of extra minutes in the response time can be the difference between life or death or a family being able to return to their home or having to sift through the ruins. Service time and service distance are constraints that can be applied to the set covering problem, whereby you're minimizing the number of locations that service the demand points within a specified period or service distance.

Before we dive into the mathematics behind the LSCP, let's take a look at a simple diagram of the problem. *Figure 10.1* depicts a set of demand points and a set of facilities from which the demand points can be serviced. We'll continue with the example from the previous paragraph and let the demand points be houses that are on fire and the facilities be fire stations. In this example, a fire station needs to be able to respond to all of the houses that are on fire within 5 minutes; otherwise, the homes will be burned to the ground. You could set up an LSCP to see if the two fire stations in this example are enough to service all of the demand. If the fire stations cannot service the demand, then there may be a need to add another fire station to this community.

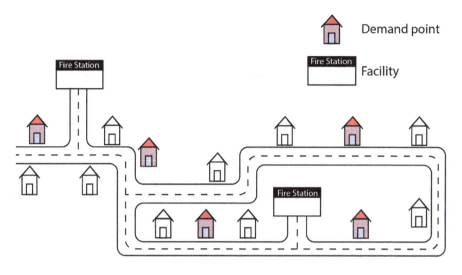

Figure 10.1 – LSCP fire example diagram

Now that we've gone over a quick example with the help of a visual aid, let's take a look at the mathematics that underpin LSCP.

Understanding the math behind the LSCP

Consider a set of service locations, $X_n = \{x_1, x_2, x_3, \dots, x_n\}$, and a set of demand points, $Y_m = \{y_1, y_2, y_3, \dots, y_n\}$. Each of these sets is located on a network, such as a street network. Let's assume that the points must be serviced within a maximum distance, S. All the points, y_i, within Y_m are then checked to see if they meet this maximum distance threshold concerning all points so that x_j within X_i so that y_i covers x_j if $d(y_i x_j) \leq S$. This can be rewritten as follows:

- Objective function: Minimize $\sum_{j=1}^{n} x_j$
- Constrained by the following:
  - $\sum_{j \in N_i} x_j \geq 1 \, \forall_i$
  - $x_j \in 0,1 \, \forall_j$

Here, we have the following:

- i is the index referencing the demand
- j is the index referencing the facility sites that can service the demand
- S is the maximum allowable service distance or time
- d_{ij} is the distance between point i and j

- $N_i = \{j \,|\, d_{ij} < S\}$
- $X_j = \begin{cases} 1, \text{if a facility is located a node } j \\ \quad\quad 0, \text{otherwise} \end{cases}$

Now that you have a better understanding of the math behind LSCPs, let's transition to a hands-on exercise where you'll solve one of these problems.

Solving LSCPs

For this example, you will be solving an LSCP using the spopt module within PySal. spopt is a collection of methods for solving spatial optimization problems such as LSCP, regionalization, and facility location problems. Let's assume that you have 150 patients who need medical attention and four medical centers from which you can serve them. The maximum service area for each hospital is 5,500 meters. To solve this LSCP, you must work through the steps outlined here:

1. Import the required packages:

    ```
    # Importing requisite packages
    from spopt.locate.coverage import LSCP
    from spopt.locate.util import simulated_geo_points

    import numpy as np
    import geopandas as gpd
    import pulp
    import spaghetti
    from shapely.geometry import Point
    import matplotlib.pyplot as plt
    import osmnx as ox
    ```

2. Define your problem space:

    ```
    patients = 150 # number of demand points represented as
    patients
    medical_centers = 4 # number of service points
    represented as medical centers
    service_area = 5500 # the max service area in meters;
    roughly 3.4 mi.
    # Setting the random seeds for reproducibility
    patient_seed = 54321
    medical_centers_seed = 54321
    # Setting up the solver
    solver = pulp.PULP_CBC_CMD(msg=False, warmStart=True)
    ```

The last section of code in the previous snippet calls upon the PuLP package, which is a **linear programming** package used to solve optimization problems. Linear programming is a mathematical technique where a linear function is maximized or minimized while being subjected to various constraints. In this example, you won't be required to code the linear function and constraints manually because they are built into PySAL's `lscp_from_cost_matrix` function, which you'll call later on in this exercise. In the next exercise, you'll have the opportunity to hand-code a linear programming function.

Jumping back to the current example, you'll be using a subset of the street network for Washington, DC. The street network has been pulled from Open Street Maps using the OSMNX package. Data from Open Street Map is licensed under the Open Data Commons Open Database License; more information can be found at `https://www.openstreetmap.org/copyright`.

The street network must be subset for this example due to the memory requirements of storing the entirety of the Washington, DC street network within the spaghetti module. In the future, it may be possible to work with larger networks in the spaghetti module as there is an ongoing initiative by the developers to refactor the code base in hopes of dramatically reducing runtimes.

3. Define the street network:

```
# Extracting the street network for Washington, DC from
Open Street Map
G = ox.graph_from_place('Washington, DC', network_
type='drive')


# Converting the network to a geopandas datafram of
edges/lines and nodes/points
gdf_nodes, gdf_edges = ox.graph_to_gdfs(G)
gdf_edges.reset_index(inplace=True)
```

After you've extracted the street network, some additional setup must be done to subset it and reproject it. The code to do this is included in the Jupyter notebook but has been left out of this section for brevity.

Now, you need to convert the street network subset into a spaghetti network, as well as create a buffered version of the street network, which you'll use later in this exercise.

4. Convert the network from a GeoPandas network into a spaghetti network:

```
ntw = spaghetti.Network(in_data=gdf_edges_clipped_p)
# Converting the spaghetti network into a new
GeoDataFrame
streets_gpd = spaghetti.element_as_gdf(ntw, arcs=True)
# Creating a buffered version of the street network for
use later
```

```
street_buffer = gpd.GeoDataFrame(
    gpd.GeoSeries(streets_gpd['geometry'].buffer(10).
unary_union),
    crs=streets_gpd.crs,
    columns=['geometry']
)
# Plotting the buffered road network
street_buffer.plot()
```

Figure 10.2 shows the theoretical street network:

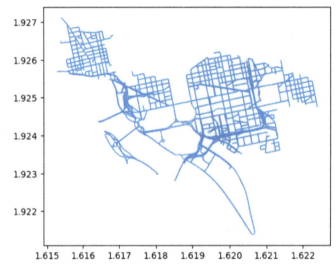

Figure 10.2 – Subset of the Washington, DC street network

Next, you'll need to leverage a utility provided by PySal called `simulate_geo_points` to simulate patient locations and medical center locations.

5. Simulate the patient and medical center locations:

```
# Simulating the patient locations
patient_locs = simulated_geo_points(street_buffer,
                                    needed=patients,
                                    seed=patient_seed)
# Simulating the medical center locations
medical_center_locs = simulated_geo_points(street_
buffer,                                    needed=medical_
centers,                                    seed=medical_
centers_seed)
```

Next, plot the patient and medical center locations onto a map with the street network. We'll symbolize the patients as green dots and the medical centers as blue crosses.

6. Map the simulated locations:

```
# Plotting the simulated points on the network
fig, ax = plt.subplots(figsize=(10,10))

streets_gpd.plot(ax=ax, alpha=0.5, zorder=1,
label='Street Grid')
patient_locs.plot(ax=ax,
color='green',zorder=2,label='Patients needing care
($n=$150)')
medical_center_locs.plot(ax=ax, markersize = 100,
color='blue',marker="P",zorder=3,label='Medical Centers
($n=$4)')
plt.legend(loc='upper right', bbox_to_anchor=(0,1))
```

Figure 10.3 shows the map produced by executing the prior code block:

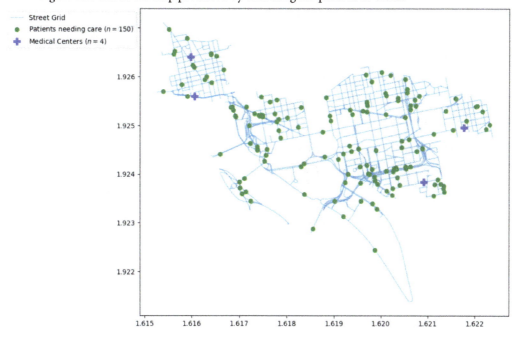

Figure 10.3 – Simulated patient and medical center locations

At times, the simulated points will not fall perfectly onto the street network. To ensure that all of the points are traversable, you need to associate the points with the network through a process known as **snapping**. Snapping moves the simulated point (here, the points represent patients and medical centers) to the closest node on the network. You'll perform this process in the next step.

7. Snap the locations to the network:

```
# Snapping the patient locations
ntw.snapobservations(patient_
locs,"patients",attribute=True)
patients_snapped = spaghetti.element_as_gdf(
    ntw, pp_name="patients", snapped = True
)


# Snapping the medical center locations
ntw.snapobservations(medical_center_locs,"medical_
centers",attribute=True)
medical_centers_snapped = spaghetti.element_as_gdf(
    ntw, pp_name="medical_centers", snapped = True
)
```

To ensure that the snapping process was performed properly, let's plot the snapped locations onto a map of the street network.

8. Plot the snapped locations:

```
# Plotting the simulated points on the network
fig, ax = plt.subplots(figsize=(10,10))

streets_gpd.plot(ax=ax, alpha=0.7, zorder=1,
label='Street Grid')
patients_snapped.plot(ax=ax,
color='green',zorder=2,label='Patients needing care
($n=$150)')
medical_centers_snapped.plot(ax=ax, markersize = 100,
color='blue',marker="P",zorder=3,label='Medical Centers
($n=$4)')
plt.legend(loc='upper right', bbox_to_anchor=(0,1))
```

Figure 10.4 shows a map of the snapped locations:

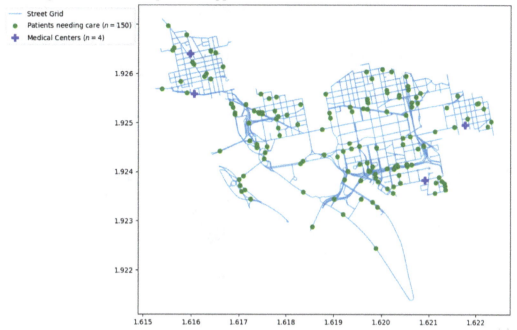

Figure 10.4 – Snapped locations

With the snapping process performed, you'll notice that the snapped locations are in line with the street network. You can use these points to calculate a distance-based cost matrix showing the distance between each patient and medical center location.

9. Calculate the distance-cost matrix:

```
cost_matrix = ntw.allneighbordistances
    ( sourcepattern=ntw.pointpatterns["patients"],
    destpattern=ntw.pointpatterns["medical_centers"], )
```

With the cost matrix calculated, you can now pass it to the PuLP solver to identify the optimal medical locations to serve all of the patient demand.

10. Find the solution to the LSCP using PuLP:

```
lscp_from_cost_matrix = LSCP.from_cost_matrix(cost_
matrix, service_area)
lscp_from_cost_matrix = lscp_from_cost_matrix.
solve(solver)
```

```
# Convert to facility client array
lscp_from_cost_matrix.facility_client_array()
```

Then, you can plot the results to identify which medical centers were selected and what part of the population they are serving. The code to plot this map is included in the Jupyter notebook but has been left out of this section for brevity.

Figure 10.5 shows the solution to the LSCP:

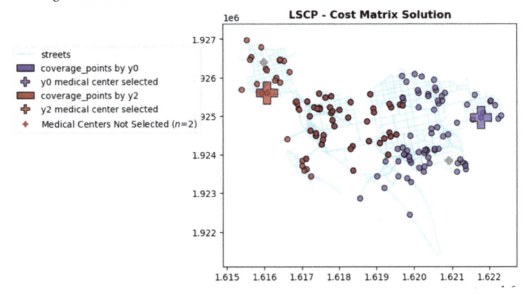

Figure 10.5 – LSCP solution

From *Figure 10.4*, you'll notice that two medical centers can service all of the patient locations in this community, with the red medical center serving the community on the left and the purple medical center serving the community on the right. Visually, it appears as if both medical centers are serving roughly the same number of people. The two medical centers that are not selected are shaded in light gray. Based on this result, there may be an opportunity to consolidate two of the medical centers into the other two centers as they are not needed to service this community.

There are many other types of coverage problems that spatial optimization can help you solve; there are too many for us to include in this chapter. Thankfully, the developers behind Spopt and PySal have created several useful tutorials that can help you continue your learning outside of this book. To access the tutorials, visit https://pysal.org/spopt/tutorials.html. For now, we'll transition to another type of spatial optimization: route-based combinatorial optimization.

Exploring route-based combinatorial optimization problems

Route-based combinatorial optimization problems set out to solve the most efficient route from point A to a set of destination points. These points are also referred to as **nodes** using terminology from **graph theory**. These problems can be simple, where a person or a vehicle departs a starting point and must travel to a set of destination points while visiting each point only once before returning to the origin point, all while minimizing the distance traveled. This is formally known as the **Traveling Salesperson Problem** (**TSP**). This problem can easily become more complicated when you add in a set of vehicles or people visiting destinations instead of a single vehicle or person in the TSP problem. This is known as a **Vehicle Routing Problem** (**VRP**). To solve this problem, you must find the optimal routes for a set of vehicles to traverse to visit a given set of customers. Additional complexity can be added to this problem class in instances where the vehicles have a set capacity that can be loaded into them. This is known as a **Capacitated Vehicle Routing Problem** (**CVRP**). We'll explore each of these problem types in more detail throughout this section, starting with the TSP.

Understanding the math behind the TSP

There are two popular **integer linear programming** formulations for the TSP. Both formulations start similarly, but they differ in how they represent the constraint that a tour can only visit each city once. The formulations start as follows:

- Minimize: $\sum_{i=1}^{n} \sum_{j \neq i, j=1}^{n} c_{ij} x_{ij}$

- Constrained by the following:

  - $\sum_{i=1, i \neq j}^{n} x_{ij} = 1 \ for \ j = 1, \dots, n$

  - $\sum_{j=1, j \neq i}^{n} x_{ij} = 1 \ for \ i = 1, \dots, n$

- Here, we have the following:

  - $x_{ij} = \begin{cases} 1, if \ a \ path \ goes \ from \ point \ i \ to \ point \ j \\ 0, otherwise \end{cases}$

As this formulation is written right now, it would allow for violations of the global constraint for the TSP: that the route must visit each destination point only once. To solve this, Miller, Tucker, and Zemlin created the **Miller-Tucker-Zemlin** (**MTZ**) formulation. The MTZ formulation adds an extra variable to the problem called u_i, which gets a value for each destination except for the origin/depot destination. As the vehicle travels from stop to stop, u_j must be bigger than u_i. This formulation eliminates the possibility of subtours that would have previously visited stops by adding the following additional constraint:

$$u_i - u_j + (n-1)x_{ij} \geq n - 2 \; for \; 2 \leq i \neq j \leq n$$

$$1 \leq u_i \leq n - 1 \; for \; 2 \leq i \leq n$$

To help you understand this concept a bit more, see *Figure 10.6* and *Figure 10.7*. In *Figure 10.6*, a delivery truck can be seen traversing between points i and j. In this depiction, node i can be thought of as the depot from which the truck is making its deliveries. Node j can be thought of as the first stop along the route:

Figure 10.6 – Truck drives from node i to node j

The vehicle conducts its route by going to node *2* through to node *6*, as depicted in *Figure 10.7*:

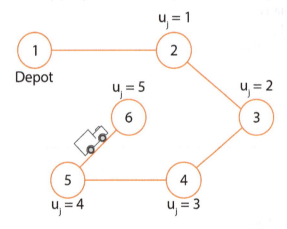

Figure 10.7 – Vehicle at node 6

When leaving node 6, the vehicle can't go back to node 2 on its way to the depot. This is because u_j for node 6 is 5 and u_i for node 2 is 1 and the constraint holds that u_i cannot be bigger than u_j. Because of this, the vehicle must return to the depot as there are no destinations left to visit.

In contrast, the **Danzig-Fulkerson-Johnson (DFJ)** formulation labels each city from 1, ..., n and utilizes subsets to eliminate the possibility of subtours. Take, for example, the route depicted in *Figure 10.8*, which depicts a route with a subtour:

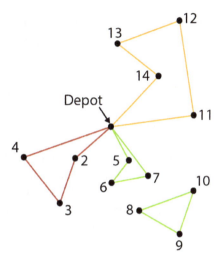

Figure 10.8 – Route with a subtour

Here, the subtour is between nodes 8, 9, and 10. The DFJ formula compares the number of arcs, or lines, between the nodes to the number of nodes within each subset. If the number of arcs equals the number of nodes, then the route is a subtour and violates the global constraint of the TSP. To correct this, one of the arcs must be removed; then, the nodes are connected to a node outside of the subset, as displayed in *Figure 10.9*:

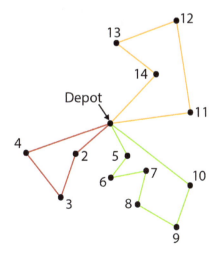

Figure 10.9 – Route with its subtour removed

The constraint that DFJ adds is as follows:

$$\sum_{i \in Q} \sum_{j \neq i, j \in Q} x_{ij} < |Q| - 1 \; for \; \forall Q \subsetneq \{1, \dots, n\}, |Q| \geq 2$$

Now that you know the math behind the TSP, let's start solving the problem with Python.

Setting up the Google Maps API

Before we begin solving the TSP, we need to introduce two new Python packages: gmap and googlemaps. As you may be able to tell from their names, these packages allow Python to interoperate with Google Maps. In this case study, you'll be leveraging the Google Maps API to calculate a distance matrix between your origin and destination points while leveraging the real-world street network within Google Maps. To leverage these packages, you must sign up for an API key with Google by visiting https://developers.google.com/maps and clicking **Get Started**. Google Maps provides you with a $200 credit each month, which can be used to pull information from their API. Once these credits have been utilized, you will begin being charged for the additional information until the next month. This case study, when run end to end, will not require you to go over your credit allocation.

Once you've set up your account, you'll need to create a new project. We've called ours **VRP Project**. Select this project and scroll down to **Enabled APIs**. Then, select **Directions API** and click **Enable**. Lastly, select the **Credentials** tab on the left-hand side and select **SHOW KEY** to reveal your API key. Copy and paste this into the code for the case study.

Solving the TSP

This case study imagines that you have 15 customers that you need to deliver products to in the most efficient manner. You'll be delivering these products to your customers from your warehouse in New York City, located at (40.749587, -73.985441). Work through the steps outlined for this TSP:

1. Import the required packages:

    ```
    import numpy as np
    import pandas as pd
    import pulp
    import itertools
    import gmaps
    import googlemaps
    import ortools
    import matplotlib.pyplot as plt
    ```

2. Set your Google Maps API key:

    ```
    API_KEY = 'YOUR API KEY'
    ```

3. Load your API key into the Google Maps `Client`:

```
gmaps.configure(api_key=API_KEY)
from googlemaps import Client
client = Client(key=API_KEY)
```

With your packages imported and your API key set up, you can now start building out the problem by simulating data that you'll base your TSP solution on. If you have real-world customer locations, then you'd use those instead of the simulated data.

4. Set some problem parameters and simulate the data:

```
# Setting the random seed for reproducibility
np.random.seed(seed=12345)
# Setting up the problem
customers = 15
# Setting the location of the warehouse
wh_lat = 40.749587
wh_lon = -73.985441
# Creating a synthetic dataset with demand points around
the warehouse location
locs = pd.DataFrame({'latitude': np.random.normal(wh_lat,
.008, customers),

                                'longitude': np.random.
normal(wh_lon, .008, customers)
                     })
# Setting the warehouse as the 0th location
cols = ['latitude','longitude']
wh = pd.DataFrame([[wh_lat,wh_lon]], columns=cols)
# Creating a final dataset
data = wh.append(locs)
# Resetting and dropping the index
data.reset_index(inplace=True)
data.drop(['index'], axis=1, inplace=True)
```

Note that we set the warehouse to the 0th index. This is a critical step as the solution will not be accurate if the warehouse is in another index position. Next, let's add a column with labels for the warehouse and the destination points and another field to store the desired color to enhance the maps that you'll produce in *step 6*.

5. Enhance the DataFrame with labels and color fields:

```
# Adding in labels
data.reset_index(inplace=True)
data.rename(columns={'index':'Label'}, inplace=True)
data['Label'] = data['Label'].astype(str)
data.at[0,'Label']='Warehouse'
# Adding in colors
data['colors'] = np.where(data['Label']=='Warehouse',
"darkslateblue", "forestgreen")
```

With the data simulated and the DataFrame enhanced, you can convert the DataFrame from pandas into GeoPandas and then map the simulated customer locations and warehouse.

6. Produce a map of the simulated data:

```
import geopandas
import contextily
data_gdf = geopandas.GeoDataFrame(
    data, geometry=geopandas.points_from_xy(data.
longitude, data.latitude, crs='EPSG:4326'))
f, ax = plt.subplots(1, figsize=(10, 10))
data_gdf.plot(ax=ax,color=data_gdf['colors'])
# Add basemap
contextily.add_basemap(
    ax,
    crs=data_gdf.crs,
    source=contextily.providers.Stamen.Watercolor,
    zoom=16
)
for x, y, label in zip(data_gdf.geometry.x, data_gdf.
geometry.y, data_gdf.Label):
    ax.annotate(label, xy=(x, y), xytext=(3, 3),
textcoords="offset points")
plt.show()
```

Figure 10.10 depicts the warehouse in slate blue with the destination points shaded in forest green:

Figure 10.10 – Simulated customer points around the warehouse

The next step of the process is to call the Google Maps Directions API to produce an **Origin-Destination Cost Matrix (O-D Cost Matrix)**. An O-D Cost Matrix shows the distance between each point and all other points. The diagonal of the matrix is 0 as the distance between a point and itself will always be 0. *Figure 10.11* shows the general structure of an O-D Cost Matrix:

$$
\begin{matrix}
0 & .5 & .7 \\
.5 & 0 & .8 \\
.7 & .8 & 0
\end{matrix}
$$

Figure 10.11 – O-D Cost Matrix example

7. Calculate the distance between the warehouse and customers:

```
import sys
np.set_printoptions(threshold=sys.maxsize)
np.set_printoptions(linewidth=1000)
# Calculating the distances from Google Maps
distances = np.zeros((len(data_gdf),len(data_gdf)))
data_gdf['coord'] = '0'
```

```
for row in range(len(data_gdf)):
    data_gdf.at[row,'coord'] = str(data_gdf.
latitude[row]) + "," + str(data_gdf.longitude[row])
for lat in range(len(data_gdf)):
    for lon in range(len(data_gdf)):
        # Call google maps api to calculate distances
        maps_api_result = client.
directions(data_gdf['coord'].iloc[
lat],                                             data_
gdf['coord'].iloc[lon],

                                       mode='driving')
        # append the distance to the distances df
        distances[lat][lon] = maps_api_result[0]['legs']
[0]['distance']['value']
# Converting array from float to int
dist_int = distances.astype(int)
# Display the distance array
print(np.matrix(distances))
```

After running the prior code cell, the O-D Cost Matrix will be displayed within the notebook. The output is shown in *Figure 10.12*:

```
[[    0.  1314.  2022.  1484.  1673.  2812.  2139.   847.  9248.  1861.  1729.  1156.  2189.   951.  1034.  2095.]
 [  714.     0.  2739.  1806.   677.  3696.  2915.  1094.  8252.  2746.  2443.  1875.  1193.  1547.   967.  2347.]
 [ 2858.  3392.     0.  2226.  3755.  2701.  2412.  2925. 11337.  1750.  2712.  3236.  3305.  3032.  3112.  4665.]
 [ 1120.  1503.  2656.     0.  1858.  3781.  3492.  1500.  9440.  2830.  2848.  2280.  1096.  1953.  1372.  2753.]
 [ 1492.  1091.  3519.  2582.     0.  4299.  3618.  1873.  8328.  3354.  3226.  2657.  1828.  2643.  1748.  3126.]
 [ 2715.  2930.  2475.  3051.  3610.     0.   666.  2467.  9254.  2387.  1285.  2275.  4123.  2261.  2654.  1475.]
 [ 2051.  2270.  2524.  2366.  2947.  1386.     0.  1805.  9546.  2436.  1008.  1611.  3462.  1597.  1990.  1766.]
 [  850.   467.  2873.  1947.  1144.  2897.  2222.     0.  8716.  2398.  1812.  1244.  1660.   917.   792.  1717.]
 [ 8713.  8052. 10731.  9805.  8245.  9842.  9905.  8836.     0. 10315.  9713.  9145.  9153.  9289.  8709.  8030.]
 [ 1423.  1642.  1884.  1736.  2319.  2047.  1373.  1175.  9890.     0.   963.  1485.  2835.  1281.  1362.  2426.]
 [ 2196.  2411.  1957.  2530.  3091.  1141.   852.  1947. 10394.  1470.     0.  1288.  3607.  1740.  2134.  3105.]
 [ 1612.  1831.  2469.  1925.  2508.  1653.   978.  1364.  9811.  1987.   568.     0.  3024.  1157.  1551.  2031.]
 [ 1665.  1259.  3431.  1475.  1475.  4556.  3801.  2046.  9055.  3605.  3393.  2825.     0.  2498.  1917.  3295.]
 [ 1048.   665.  3103.  2145.  1342.  2781.  2106.   803.  8912.  2596.  1696.  1128.  1858.     0.   990.  1601.]
 [  878.  1061.  2081.  1544.  1772.  2557.  1884.   594.  9347.  1606.  1474.   903.  2288.   698.     0.  1842.]
 [ 2236.  1776.  4383.  3332.  2453.  2286.  2244.  1780.  7769.  3783.  2156.  1588.  2969.  1574.  2177.     0.]]
```

Figure 10.12 – O-D Cost Matrix output

The next step is to define the combinatorial optimization problem using the **PuLP** package within Python. PuLP is a package that was built for optimization using integer programming. As we discussed earlier in this section, there are two possible formulations for the TSP. For this example, you'll leverage the MTZ formulation. You'll leverage the DFJ method in the next example.

8. Use PuLP to write the optimization problem:

```
# Set the problem
tsp_problem = pulp.LpProblem('tsp_mip', pulp.LpMinimize)
# Defining the problem variables
x = pulp.LpVariable.dicts('x', ((i, j) for i in
range(customers+1) for j in range(customers+1)), lowBound
= 0, upBound = 1, cat='Binary')
# Tracking the order the points are serviced to prevent
subtours
u = pulp.LpVariable.dicts('u', (i for i in
range(customers+1)), lowBound = 1, upBound =
(customers+1), cat='Integer')
# Establishing the objective function
tsp_problem += pulp.lpSum(dist_int[i][j] * x[i,j] for i
in range(customers+1) for j in range (customers+1))
```

9. Program the constraints:

```
# Constraint 1
for i in range(customers+1):
    tsp_problem += x[i, i] == 0
# Constraint 2
for i in range(customers+1):
    tsp_problem += pulp.lpSum(x[i, j] for j in
range(customers+1)) == 1
    tsp_problem += pulp.lpSum(x[j, i] for j in
range(customers+1)) == 1
# Constraint 3 - Eliminates the possibility of a subtour
using MTZ formulation
for i in range(customers+1):
    for j in range(customers+1):
        if i != j and (i != 0 and j != 0):
            tsp_problem += u[i] - u[j] <= (customers+1) *
(1 - x[i, j]) - 1
```

With the problem variables, objective function, and constraints written, you can now solve the TSP. The following code block will solve the TSP. If PuLP can find an optimal solution, then it will print the optimal solution and the distance traveled. If PuLP is unable to find an optimal solution, it will write an error message under the Jupyter notebook cell.

10. Solve the TSP:

```
# Solve the TSP
status = tsp_problem.solve()
status, pulp.LpStatus[status], pulp.value(tsp_problem.
objective)
```

The status of the PuLP solver and the distance traveled by the route should now be displayed in your notebook and look something like (1, 'Optimal', 32585.0). PuLP was able to find an optimal solution that traveled 32,585 meters or approximately 20.25 miles. Depending on the time of day in which you ran this code, the Google Maps API may have returned different distance calculations between points, which will impact the distance of the optimal route. Now, let's plot the results to see the order in which the customers were visited.

11. Produce a map of the optimal route:

```
f, ax = plt.subplots(1, figsize=(10, 10))
data_gdf.plot(ax=ax,color=data_gdf['colors'])
# Add basemap
contextily.add_basemap(
    ax,
    crs=data_gdf.crs,
    source=contextily.providers.Stamen.Watercolor,
    zoom=16
)
for lon, lat, label in zip(data_gdf.geometry.x, data_gdf.
geometry.y, data_gdf.Label):
    ax.annotate(label, xy=(lon, lat), xytext=(3, 3),
textcoords="offset points")
# Plot the optimal route between stops
routes = [(i, j) for i in range(customers+1) for j in
range(customers+1) if pulp.value(x[i, j]) == 1]
arrowprops = dict(arrowstyle='->',
connectionstyle='arc3', edgecolor='darkblue')
for i, j in routes:
    ax.annotate('', xy=[data_gdf.iloc[j].geometry.x,
data_gdf.iloc[j].geometry.y], xytext=[data_gdf.
iloc[i].geometry.x, data_gdf.iloc[i].geometry.y],
arrowprops=arrowprops)
plt.show()
```

Figure 10.13 displays the optimal route from the warehouse to the 15 customers with arrows, indicating the direction of travel from point to point:

Figure 10.13 – TSP optimal path

Now that you have a better understanding of TSPs, let's discuss **Vehicle Routing Problems** (VRPs).

Exploring a single-vehicle Vehicle Routing Problem (VRP)

VRPs are very similar to TSPs. If you have a single vehicle being routed to a set of destinations, it is identical to the TSP. To prove this point, let's solve a single-vehicle VRP. This will help you understand the context regarding how problem types can be called different things but produce similar results. This also allows you to use the alternative DFJ formulation mentioned earlier.

Let's start by adding a few new problem parameters. We'll set the number of vehicles to 1 and also simulate some data, indicating the number of products each customer wants. We'll call this new variable demand:

1. Simulate demand and set the number of vehicles:

    ```
    # Additional Problem Parameters
    vehicles = 1
    # Adding simulated demand to the dataset
    ```

```
demand = np.random.randint(2,12,customers).tolist()
demand = [0] + demand
data_gdf['customer_demand'] = demand
```

Next, you must set up the problem by adding another dimension to the matrix, *k*, for the vehicles being used to service customer demand. For this case study, we'll use the DFJ formulation to prevent subtours, which contrasts with the MTZ formulation used in the previous case study.

2.   Set up the problem:

```
# Setting up the PuLP Solver
for vehicles in range(1, vehicles+1):
    # Linear Programming Problem
    lp_problem = pulp.LpProblem("VRP", pulp.LpMinimize)
    # Defining problem variables which are binary
    x = [[[pulp.LpVariable("x%s_%s,%s"%(i,j,k),
cat="Binary") if i != j else None
            for k in range(vehicles)]
          for j in range(customers+1)]
        for i in range(customers+1)]
    # Setting the objective function
    lp_problem += pulp.lpSum(dist_int[i][j] * x[i][j][k]
if i != j else 0
                             for k in range(vehicles)
                             for j in range(customers+1)
                             for i in range(customers+1))
```

3.   Program the constraints:

```
    # Adding in the constraints
    for j in range (1, customers+1):
        lp_problem += pulp.lpSum(x[i][j][k] if i
!= j else 0 for i in range(customers+1) for k in
range(vehicles)) == 1
    for k in range(vehicles):
        lp_problem += pulp.lpSum(x[0][j][k] for j in
range(1, customers+1)) == 1
        lp_problem += pulp.lpSum(x[i][0][k] for i in
range(1, customers+1)) == 1
    for k in range(vehicles):
        for j in range(customers+1):
```

```
        lp_problem += pulp.lpSum(x[i][j][k] if i != j
else 0
                                for i in
range(customers+1)) - pulp.lpSum(x[j][i][k] for i in
range(customers+1)) == 0
```

4. Add another constraint to prevent subtours using the DFJ formulation:

```
    subtours = []
    for i in range(2, customers+1):
        subtours += itertools.combinations(range(1,
customers+1), i)
    for s in subtours:
        lp_problem += pulp.lpSum(x[i][j][k] if i != j
else 0 for i, j in itertools.permutations(s,2) for k in
range(vehicles)) <= len(s) -1
    if lp_problem.solve() == 1:
        print('# Required
Vehicles:',vehicles)        print('Distance:',pulp.
value(lp_problem.objective))
        break
```

The prior code block will print the number of vehicles used and the distance traveled into your notebook. The distance traveled is 32,585 meters, which is identical to the distance from the TSP solution. Let's display the map to see if our route is the same.

5. Display the VRP solution:

```
f, ax = plt.subplots(1, figsize=(10, 10))
data_gdf.plot(ax=ax,color=data_gdf['colors'])
# Add basemap
contextily.add_basemap(
    ax,
    crs=data_gdf.crs,
    source=contextily.providers.Stamen.Watercolor,
    zoom=16
)
for lon, lat, label in zip(data_gdf.geometry.x, data_gdf.
geometry.y, data_gdf.Label):
    ax.annotate(label, xy=(lon, lat), xytext=(3, 3),
textcoords="offset points")
```

```
# Plot the optimal route between stops
routes = [(k, i, j) for k in range(vehicles) for i in
range(customers+1) for j in range(customers+1) if i != j
and pulp.value(x[i][j][k]) == 1]
arrowprops = dict(arrowstyle='->',
connectionstyle='arc3', edgecolor='darkblue')
for k, i, j in routes:
    ax.annotate('', xy=[data_gdf.iloc[j].geometry.x,
data_gdf.iloc[j].geometry.y], xytext=[data_gdf.
iloc[i].geometry.x, data_gdf.iloc[i].geometry.y],
arrowprops=arrowprops)
plt.show()
```

Figure 10.14 maps our optimal solution for the VRP problem:

Figure 10.14 – Single-vehicle VRP solution

The route traveled by our single-vehicle VRP is indeed identical to that of the TSP, even when we use different formulations. However, this is not entirely realistic as the vehicle is delivering a ton of packages to these customers. To get the total number of packages demanded, you must execute the following code snippet.

6. Calculate the total demand:

```
# Count the total amount of demand
data_gdf['customer_demand'].sum()
```

Our customers are demanding 86 packages from us in total, with each customer needing a different amount. Now, let's assume that our vehicle can only hold 40 packages at a time. This capacity constraint has now turned our simple VRP into a **Capacitated Vehicle Routing Problem (CVRP)**. You'll solve variation in the next case study.

Exploring a Capacitated Vehicle Routing Problem (CVRP)

At the end of the previous example, you learned that your vehicle can only hold 40 packages at a time, but your customers want 86 packages delivered. You'll now need to not only find the optimal path to deliver these packages but also the optimal number of vehicles needed to service this demand. Let's assume that your company has five vehicles in total, each with an identical capacity of 40 packages. Now, let's work through the CVRP:

1. Add additional parameters for capacity and the number of vehicles available:

```
# Additional Problem Parameters
vehicles = 5
capacity = 40
```

2. Set up the problem by adding another constraint for the capacity of each vehicle:

```
# Setting up the PuLP Solver
for vehicles in range(1, vehicles+1):
    # Linear Programming Problem
    lp_problem = pulp.LpProblem("CVRP", pulp.LpMinimize)
    # Defining problem variables which are binary
    x = [[[pulp.LpVariable("x%s_%s,%s"%(i,j,k),
cat="Binary") if i != j else None
            for k in range(vehicles)]
          for j in range(customers+1)]
        for i in range(customers+1)]
    # Setting the objective function
    lp_problem += pulp.lpSum(dist_int[i][j] * x[i][j][k]
if i != j else 0
                             for k in range(vehicles)
                             for j in range(customers+1)
                             for i in range(customers+1))
```

3. Program the constraints:

```
for j in range (1, customers+1):
    lp_problem += pulp.lpSum(x[i][j][k] if i
!= j else 0 for i in range(customers+1) for k in
range(vehicles)) == 1
for k in range(vehicles):
    lp_problem += pulp.lpSum(x[0][j][k] for j in
range(1, customers+1)) == 1
    lp_problem += pulp.lpSum(x[i][0][k] for i in
range(1, customers+1)) == 1
for k in range(vehicles):
    for j in range(customers+1):
        lp_problem += pulp.lpSum(x[i][j][k] if i != j
else 0
                                        for i in
range(customers+1)) - pulp.lpSum(x[j][i][k] for i in
range(customers+1)) == 0
for k in range(vehicles):
    lp_problem += pulp.lpSum(data.customer_demand[j]
* x[i][j][k] if i != j else 0 for i in range(customers+1)
for j in range (1,customers+1)) <= capacity
subtours = []
for i in range(2, customers+1):
    subtours += itertools.combinations(range(1,
customers+1), i)
for s in subtours:
    lp_problem += pulp.lpSum(x[i][j][k] if i != j
else 0 for i, j in itertools.permutations(s,2) for k in
range(vehicles)) <= len(s) -1
if lp_problem.solve() == 1:
    print('# Required Vehicles:',vehicles)
    print('Distance:',pulp.value(lp_problem.
objective))
    break
```

The optimal solution to the CVRP is three vehicles traveling a total distance of 34,765 meters or roughly 21.6 miles. Now, let's map the route for each vehicle in the next step. We'll plot the route of the first vehicle in blue, the second vehicle in purple, and the third vehicle in black. The code to produce the map is included in the Jupyter notebook but has been left out of this section for brevity.

Figure 10.15 shows the optimal path for the three vehicles:

Figure 10.15 – CVRP optimal paths

Vehicle 3 travels to fewer points that are much closer together. In the real world, this would have potential impacts on staff needed to operate the vehicles as the third vehicle is not leveraged as much in this scenario as the other two.

In this section, we based our results on the Google Map API. This problem can also be solved by loading a street network shapefile into PySAL's Spaghetti library to calculate the distance. To learn more about this method, visit `https://pysal.org/spaghetti/notebooks/tsp.html`.

Summary

In this chapter, we introduced you to different types of spatial optimization that you can solve using Python. We kicked this chapter off by discussing LSCPs, where you found the optimal number of facilities needed to serve the emergency service demand of a community.

Then, we transitioned our focus to discussing route-based optimization problems, including TSP, VRP, and CVRP.

You explored three different case studies in this section using two different integer linear programming formulations (MTZ and DFJ), which help the optimization abide by the global constraint that a person or vehicle can only visit each stop once.

You were also introduced to a handful of new packages, including Spopt, which is PySAL's spatial optimization library. You also learned about PuLP for solving integer optimization problems. Lastly, you set up a Google Maps API to gather real-world distances in the form of an O-D Cost Matrix from Google Map's street network.

We covered a lot of content in this chapter, but we have just scratched the surface in terms of the problems you can solve with spatial optimization. As a reminder, these problems can grow in complexity and computation time as you add more constraints that model the real world. We encourage you to leverage some of the additional resources mentioned within this chapter to continue your learning.

In *Chapter 11, Advanced Topics in Spatial Data Science*, we'll cover some additional topics that take some of the topics covered within prior chapters to an advanced level. We'll also introduce a few new topics around ethics in spatial data science.

11

Advanced Topics in Spatial Data Science

Welcome to the final chapter of this book. We've covered an incredible amount of content in the prior chapters, from introducing you to the concept of spatial data science to working on numerous challenging case studies applying cutting-edge geospatial machine learning algorithms. Unlike previous chapters, this one will not focus on a single concept but instead will explore a handful of more advanced topics that will further enhance your learnings from previous chapters. We'll also dive into the topic of ethics in data science and spatial data science, which is becoming a topic of rapidly growing interest due to the improvements in technology and data access, combined with data breaches and other dilemmas.

In this chapter, we'll cover the following topics:

- Efficient operations with spatial indexing
- Estimating unknowns with spatial interpolation
- Ethical spatial data science

Technical requirements

For this chapter, you'll leverage the Jupyter notebooks titled `Chapter 11 - Spatial Indexing` and `Chapter 11 - Spatial Interpolation`, which can both be found in the GitHub repository for this book at `https://github.com/PacktPublishing/Applied-Geospatial-Data-Science-with-Python/tree/main/Chapter11`.

Efficient operations with spatial indexing

Over the course of this book, we've worked with spatial datasets of varying sizes. However, given the nature of the case studies and the need for simplicity, we haven't worked with very large spatial datasets. As spatial datasets grow in size and cover larger geographic areas, you will often need to find ways to access and perform operations on the data more efficiently. One way to add efficiency to your spatial data science workflows is through the use of **spatial indexing**. A spatial index is a way of structuring your data in a way that makes accessing and performing operations on the spatial object more efficient as compared to sequentially scanning every record in the dataset. Spatial indexing, at times, can dramatically increase the speed of spatial operations, including spatial joins and intersections.

There are many types of spatial indexes available in both commercial and **open source software,** and there are far too many for us to discuss each at length within this chapter. For brevity, we'll talk about the **R-tree spatial index**, which is one of the most common spatial indexes in use today. An R-tree represents observations and their **minimum bounding rectangles** (**MBRs**) as the lowest level of the spatial index. The *R* in R-tree stands for *rectangle*. After constructing the lower-order MBR, the R-tree iteratively aggregates each object with other nearby objects and constructs a new aggregated MBR. This process is performed over and over again until there is a single bounding rectangle covering all objects. *Figure 11.1* shows how an R-tree is constructed:

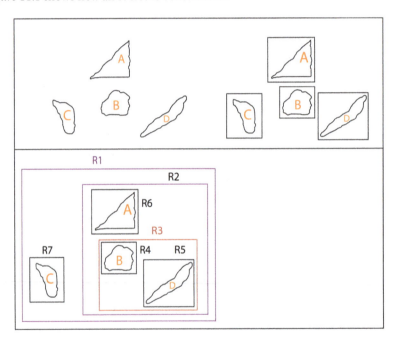

Figure 11.1 – R-tree spatial indexing

Once the R-tree is constructed, the hierarchical structure of the spatial objects can be represented, as shown in *Figure 11.2*:

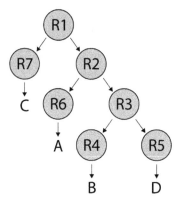

Figure 11.2 – R-tree hierarchical structure

When a spatial operation, such as an intersection, is performed on an R-tree-indexed object, the R-tree first creates a query box. Starting at the top of the tree, the R-tree uses this query box to see if any of the MBRs intersect this query box. It then expands the search to identify the intersecting child nodes. This process is performed in a recursive manner over and over again until the lowest level of the R-tree is reached. Once this process is completed, all of the intersecting objects are returned.

Now that you understand the fundamentals of the R-tree index, let's see how it can be implemented inside of GeoPandas and used to perform a spatial query.

Implementing R-tree indexing in GeoPandas

Implementing an R-tree spatial index inside of GeoPandas is quite simple. After you've loaded your dataset into a GeoPandas DataFrame, you simply execute the `.sindex` method on the GeoDataFrame. To put this into practice, let's work through a brief exercise using the Airbnb dataset you've worked with in prior chapters. We'll perform a similar spatial operation where we filter the Airbnb data down to that within Manhattan, similar to what was done in *Chapter 9, Developing Spatial Regression Models*. Follow these steps:

1. Read in the data and convert it to a GeoDataFrame:

    ```
    # Reading in the data as a Pandas DataFrame
    listings = pd.read_csv(path + r'NY Airbnb June 2020\
    listings.csv.gz', compression='gzip', low_memory=False)
    # Converting it to a GeoPandas DataFrame
    listings_gpdf = gpd.GeoDataFrame(
        listings,
        geometry=gpd.points_from_xy(listings['longitude'],
    ```

```
                                            listings['latitude'],
                                            crs="EPSG:4326")

)
```

The `listings` GeoDataFrame has 37,410 records and covers the geographic area of the state of New York.

2. Execute a filter without using a spatial index:

```
# Focusing on attractions in Manhattan, so we need to
create a mask to filter locations in the Manhattan
borough
boroughs = gpd.read_file(path + r"NYC Boroughs\nybb_22a\
nybb.shp")
boroughs = boroughs.to_crs('EPSG:4326')
manhattan = boroughs[boroughs['BoroName']=='Manhattan']
# get the start time
st = time.time()
listings_mask = listings_gpdf.within(manhattan.loc[3,
'geometry'])
listings_manhattan = listings_gpdf.loc[listings_mask]
# get the end time
et = time.time()
# get the execution time
elapsed_time = et - st
print('Execution time:', elapsed_time, 'seconds')
```

Without the spatial index, the filtering operation ran in roughly 57.06 seconds and identified 15,284 Airbnb listings in Manhattan. Now, let's add in the R-tree-based spatial index and compare the execution time.

3. Perform a filter using the spatial index:

```
# Building the R-tree spatial index
sindex = listings_gpdf.sindex
# get the start time
st = time.time()
# Getting the indexes of the possible matches
idex_possible_matches = list(sindex.
intersection(geometry.bounds))
```

```
# subsetting the dataframe to be only possible matches
possible_matches_df = listings_gpdf.iloc[idex_possible_
matches]
# Performing an intersection to get the precise matches
precise_matches_df = possible_matches_df[possible_
matches_df.intersects(geometry)]
# get the end time
et = time.time()
# get the execution time
elapsed_time = et - st
print('Execution time:', elapsed_time, 'seconds')
```

The operation now ran in roughly 52.95 seconds, saving about 5 seconds from the original runtime, and identified the same 15,284 Airbnb listings in Manhattan. Saving 5 seconds isn't that big of a saving, so you may wonder why the time saving wasn't larger. One limitation of R-tree indexes is that if the filtering polygon and spatial observations have the same MBR, the R-tree index will identify each of the points as a possible match and not yield huge efficiency gains.

One way to work around this is to subdivide your filtering polygon into smaller polygons and then iterate through those smaller polygons to identify matching locations.

4. Create a subdivided polygon from the Manhattan shapefile:

```
subdivided_polygon = ox.utils_geo._quadrat_cut_
geometry(geometry, quadrat_width=1) # quadrant_width is
in the CRS measurement units (4326:degrees)
```

5. Iterate through the subdivided polygons to identify Manhattan Airbnb listings:

```
# get the start time
st = time.time()
points_in_geometry = pd.DataFrame()
for geom in subdivided_polygon:
    # add in a slight buffer to account for points
falling on the lines of the subdivided polygons
    geom = geom.buffer(1e-14).buffer(0)
    # Getting the indexes of the possible matches from
the R-tree
    idex_possible_matches = list(sindex.
intersection(geom.bounds))
```

```
        possible_matches_df = listings_gpdf.iloc[idex_
possible_matches]
        # Performing an intersection to get the precise
matches
        precise_matches_df = possible_matches_df[possible_
matches_df.intersects(geom)]
        points_in_geometry = points_in_geometry.
append(precise_matches_df)
# get the end time
et = time.time()
# get the execution time
elapsed_time = et - st
print('Execution time:', elapsed_time, 'seconds')
```

Using this method, the execution time was reduced to roughly 16.05 seconds and once again identified the same 15,284 Airbnb listings in Manhattan. With this method, the execution time was reduced by 41.01 seconds or roughly 71%. With larger and larger datasets, identifying efficiencies such as this can save you countless hours throughout your workflow.

As we mentioned at the start of this section, multiple spatial indexes can be used in your analysis. In fact, over the last few years, new spatial indexes have been created to help solve problems related to spatial big data, which we'll discuss in the next section. One such recently developed index is H3.

Introducing the H3 spatial index

The **Hexagonical Hierarchical (H3)** index was developed by Uber to help it analyze supply and demand within its ride-sharing marketplace efficiently and to help it optimally price its services. Uber released H3 to the public under an open source license in 2018.

The H3 grid system comprises nesting hexagons that cover the Earth's surface. There are a total of 16 resolutions of the H3 grid system, with the first level containing 122 hexes and the 16th level containing over 569 trillion unique hexes. The area of the hexes in each level gets progressively smaller and is 1/7th the size of the hexes in the prior level. For each resolution, the hexes are assigned unique identifiers that can be appended to other geographic data to aid in efficient spatial analysis and feature engineering. For brevity, we won't go into more extensive detail in this chapter, and instead, we encourage you to read Uber's blog post at https://www.uber.com/blog/h3/.

Let's transition our focus to leveraging the H3 index to create newly engineered variables. As a reminder, we previously discussed variable engineering in *Chapter 7, Spatial Feature Engineering*. For this example, we'll use the H3 index to quickly count the number of Airbnb listings across Manhattan. Follow these steps:

1. Import the required packages:

```
%matplotlib inline
from h3 import h3
import contextily
```

2. Define a function to append an H3 identifier to each listing:

```
# Set the H3 resolution
h3_resolution = 8
# Creating a function to add the H3 identifier to each of
the Airbnb Points
def add_h3_id(row):
    return h3.geo_to_h3(
        row.geometry.y, row.geometry.x, h3_level)
# Executing the function
listings_manhattan['h3'] = listings_gpdf.apply(add_h3_id,
axis=1)
```

3. Summarize the listings within each hexagon:

```
# Creating a dataframe with the count of airbnb's within
each hexagon
airbnb_count = listings_manhattan.groupby(['h3']).
h3.agg('count').to_frame('count').reset_index()
```

4. Append the hexagon geometry to the aggregated DataFrame:

```
# Defining a function to get the geometry for each of the
H3 hexagons
def add_h3_geometry(row):
    points = h3.h3_to_geo_boundary(
        row['h3'], True)
    return Polygon(points)
# Adding the geometry to the airbnb_count dataframe
```

```
airbnb_count['geometry'] = airbnb_count.apply(add_
geometry, axis=1)
# Converting to a geodataframe
gdf = gpd.GeoDataFrame(counts, crs='EPSG:4326')
```

5. Plot a choropleth map of the Airbnbs within each cell:

```
f, ax = plt.subplots(1, figsize=(10, 20))
# Plot choropleth of counts
gdf.plot(
    column='count', cmap='coolwarm', scheme='quantiles',
    k=4, edgecolor='white', linewidth=0.1, alpha=0.5,
    legend=True, ax=ax
)
# Add basemap
contextily.add_basemap(ax, crs=gdf.crs,
source=contextily.providers.Stamen.TonerLite,
)
# Remove axis
ax.set_axis_off()
# Display the map
plt.show()
```

Figure 11.3 shows the resulting choropleth map of the density of Airbnbs per H3 cell in Manhattan:

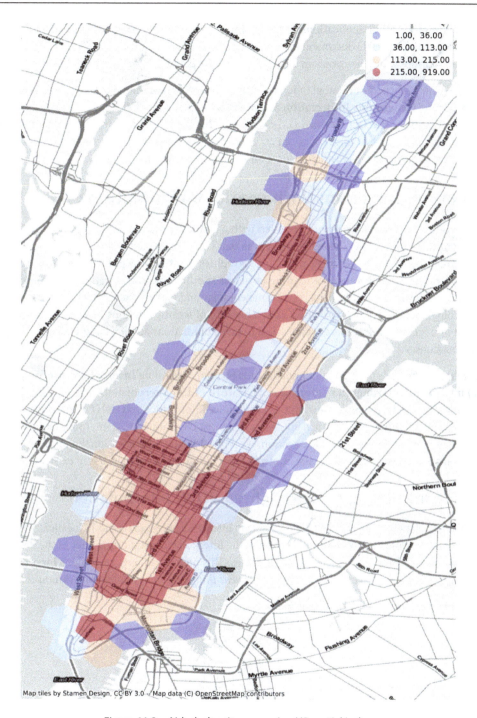

Figure 11.3 – Airbnb density map using H3 spatial index

There are numerous other applications of the H3 index, which we won't cover in this section. We encourage you to visit the H3 documentation at `https://uber.github.io/h3-py/intro.html` to learn more about how you can use H3 in your geospatial data science workflows.

To advance your knowledge of spatial indexing even further, we recommend reading the **University Consortium for Geographic Information Science's (UCGIS's)** *GIS&T Body of Knowledge*, paying particular attention to the section titled *DM-66 - Spatial Indexing*. This resource can be accessed by visiting `https://gistbok.ucgis.org/bok-topics/spatial-indexing`. For now, we'll move on to our next topic, which focuses on spatial interpolation.

Estimating unknowns with spatial interpolation

Over the course of time, you may be presented with a geospatial dataset with a sparse number of observations that do not cover the entire study area that you're interested in analyzing. As such, you may be looking for a way to fill in the missing geographies. **Spatial interpolation** is a process that uses known values from observations to estimate values at other unknown locations. This process is common in a number of scientific fields, such as meteorology and wildlife conservation. When it comes to meteorology, weather data is provided from a handful of weather stations in a given geography. From that information, meteorologists are asked to make predictions about what the weather will be at other locations.

There are many methods for performing spatial interpolation, including **Inverse Distance Weighted** (**IDW**) interpolation, **Triangular Information Network** (**TIN**) interpolation, and **Kriging**-based interpolation methods, to name a few. In this section, we'll introduce you to IDW interpolation as well as **Ordinary Kriging** (**OK**), which are two of the most common spatial interpolation methods that you'll encounter. At the end of the section, we'll provide you with additional resources where you can continue to learn more about the numerous spatial interpolation methods in existence today.

Applying Inverse Distance Weighted (IDW) interpolation

The IDW interpolation method uses sampled points to infer the value at an unknown location, whereby the effect of the sampled point on the prediction of the interpolated point diminishes as the distance between the sampled and interpolated point increases. This method adheres to **Tobler's First Law of Geography**, which states *"everything is related, but near things are more related than distant things"*.

The mathematical formula for IDW is as follows:

$$Z_j = \frac{\sum_{i=1}^{n}\left(\frac{z_i}{d_{ij}^P}\right)}{\sum_{i=1}^{n}\left(\frac{1}{d_{ij}^P}\right)}$$

Here, the following applies:

- Z_j is the value of the unknown point
- d_{ij}^p is the distance between the unknown point and the sampled point
- P is the user-selected power function
- n is the number of sampled points used to estimate the unknown point

As a quick example to demonstrate how the mathematics of IDW works, review *Figure 11.4*. In *Figure 11.4*, your goal is to interpolate the value of point x using the five nearest observations:

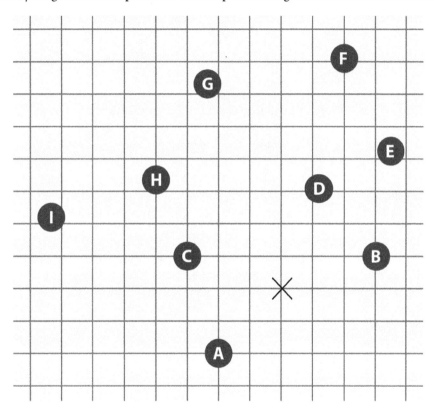

Figure 11.4 – Simple IDW interpolation example

Figure 11.4 is supplemented with *Table 11.1*, which has information about the value of each observation to point *x*:

Observation	Value	Distance
A	7	1.5
B	6	1.3
C	7.5	1.4
D	4	1.5
E	5	3.2
F	3	4.4
G	3.5	4.2
H	2	3.3
I	.5	4.6

Table 11.1 – IDW interpolation: simple example data

The five nearest observations to point *x* in order of increasing distance are *B*, *C*, *A*, *D*, and *E*. To calculate the interpolated value for point *x*, you'll first sum the value of the observation divided by the distance, which is (6/1.3)+(7/5/1.4)+(7/1.5)+(4/1.5)+(5/3.2), which is equal to roughly 18.86. Next, you'll sum up 1 divided by the distance, which is (1/1.3)+(1/1.4)+(1/1.5)+(1/1.5)+(1/3.2), which is equal to roughly 3.13. Lastly, you'll divide 18.86 by 3.13, which gives you a value of roughly 6.02 as the interpolated value of point *x*.

To perform IDW inside of Python, we'll leverage the `pyidw` Python package developed by Yahya Tamim as it is a lightweight and simple-to-use package for performing IDW interpolation. For this example, you'll be interpolating the average daily temperature in Fahrenheit for the state of Ohio. The dataset you'll leverage for this exercise was obtained from the **National Oceanic and Atmospheric Administration's (NOAA's)** National Centers for Environmental Information Climate Data Online portal at `https://www.ncei.noaa.gov/cdo-web/datasets`. You'll be interpolating the temperatures for January 1, 2022. Follow these steps:

1. Read in the data file:

```
# Ohio Shapefile
ohio = gpd.read_file(path + "Ohio\Ohio.shp")
ohio_state = ohio.dissolve()
ohio_state = ohio_state[['STATEFP','geometry']]
ohio_state = ohio_state.to_crs("EPSG:4326")
ohio_state.to_file(path + "Ohio\Ohio_State.shp")
```

```
# Ohio Temperatures
temps = pd.read_csv(path + "Chapter 11\Ohio Temps.csv")
temps = temps[~temps['TAVG'].isnull()]
temps = temps[temps['DATE']=='2022-01-01']
temps = temps[['LATITUDE','LONGITUDE','TAVG']]
# Converting to geopandas and writing out as shapefile
temps_gpdf = gpd.GeoDataFrame(
    temps,
    geometry=gpd.points_from_xy(temps['LONGITUDE'],
                                temps['LATITUDE'],
                                crs="EPSG:4326")
)
# Output the shapefile for use later on
temps_gpdf.to_file(path + "Chapter 11\Ohio Temps.shp")
Explore the temperature observations
```

2. Format the dataset and plot the observations:

```
# Dropping the geometry column
temps_nogeom = temps.drop(['geometry'], axis=1)


# Converting the DataFrame to a numpy array
temps_array = temps_nogeom.to_numpy()
# Plot the observations
obs = plt.scatter(temps_array[:, 1],
    temps_array[:, 0],
    c = temps_array[:, 2], cmap='coolwarm')
cbar = plt.colorbar(obs)
plt.title('Observed Temperatures')
plt.show()
```

Figure 11.5 shows a scatter plot of the observed temperatures for January 1, 2022:

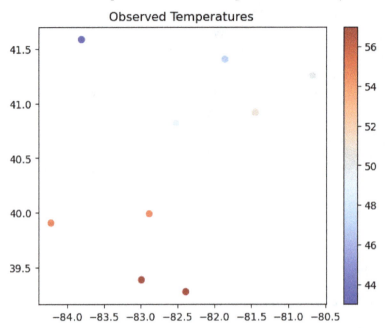

Figure 11.5 – Average daily temperature for Ohio on January 1, 2022

Through analysis of the scatter plot, you can deduce that there are nine weather stations reporting data for January 1, 2022 across Ohio. The highest average daily temperatures are observed in the southern portions of the state, where they average around 56 °F.

In the next step, you'll perform IDW interpolation using this data to estimate the temperatures for the entire state of Ohio. To do this, you'll call the `idw.idw_interpolation` function, passing to it the following:

- A shapefile of the observations
- A shapefile for the geographic area in which IDW will be performed; here, it is the state of Ohio
- The column storing the value to be interpolated
- The chosen power
- The search radius, or the number of points to be considered
- The output resolution for the raster of interpolated values

3. Perform IDW interpolation:

```
idw.idw_interpolation(
    input_point_shapefile="path + Chapter 11\Ohio Temps.
shp",
    extent_shapefile="path + Ohio\Ohio_State.shp",
    column_name="TAVG", power=2,
    search_radious=3, output_resolution=250
)
```

Figure 11.6 shows the interpolated temperature across Ohio from the IDW method:

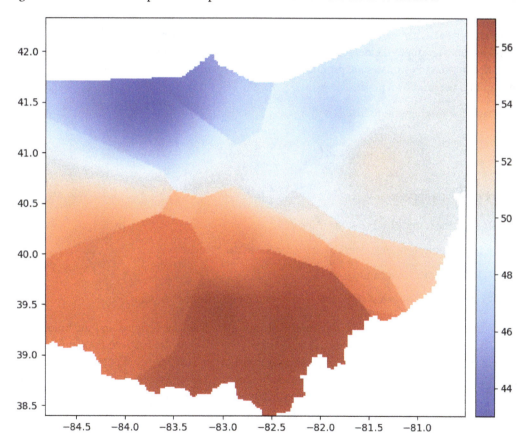

Figure 11.6 – IDW interpolated average daily temperatures

The IDW interpolation does a good job of picking up on the overall signal from the data, with temperatures decreasing as you move further north in the state. It also picks up on the observation in the northeast of the state, which is a bit warmer than other locations in the surrounding area.

The IDW method is a relatively simple method that does have a few limitations. Firstly, the resulting interpolation's accuracy can decrease if the distribution of the sampled points is unevenly distributed across the study area. Secondly, the IDW method cannot produce interpolated values that are outside the minimum or maximum of the observed values. Given these limitations, other more advanced interpolation methods have been developed. In the next section, we'll introduce you to Kriging, which is a more advanced spatial interpolation method.

Introduction to Kriging-based interpolation

Kriging is a spatial interpolation method that leverages distance similarly to the IDW interpolation method but also leverages the degree of variation in sampled data points to interpolate unknown points. Kriging-based estimates are produced using a weighted linear combination of the sampled values around the unknown point.

The formula for Kriging is as follows:

$$\hat{Z}(s_0) = \sum_{i=1}^{N} \lambda_i Z(s_i)$$

Here, the following applies:

- $Z(s_i)$ is the sampled value at the ith location
- λ_i is an unknown weight for the sampled value at the ith location
- s_0 is the prediction location
- N is the number of sampled points used to estimate the unknown point

In IDW interpolation, λ_i is based solely on the distance between the sampled and the unknown point. Conversely, in Kriging-based methods, λ_i is based on the distance between the sampled and unknown points as well as the overall spatial arrangement between the sampled points. As such, the spatial autocorrelation between the points needs to be measured using a fitted model to the observed points. The process of fitting a model is called **variography**, and the fitted model is thusly called a **semivariogram**. The formula for a semivariogram is as follows:

$$\hat{\gamma}(h) = \frac{1}{2N} \sum_{i=1}^{N} [\{Z(x_i + h) - Z(x_i)\}^2]$$

Here, the following applies:

- γ is the semivariogram
- h is the distance vector, also known as the lagged distance
- $Z(x_i)$ is the value of the sampled point at location i
- N is the number of sampled points used to estimate the unknown point

Once the semivariogram is calculated, a model is fit to the points, which is known as the empirical semivariogram. The model that is fit to the points is similar to that of a regression model, where many functions can be fit to the points. There are several variogram models that can be fit in most Kriging, applications including Gaussian, exponential, spherical, and linear. The application of each model depends on the phenomena being evaluated and will impact the interpolated values. To learn more about each of these models, we encourage you to visit `https://gisgeography.com/semi-variogram-nugget-range-sill/#mathematical-function-and-models`.

The Kriging formula explained previously is known as **OK** and is one of the most common forms of Kriging and the simplest to implement. In addition to OK, there are many other variations, including **Universal Kriging (UK)**, **Indicator Kriging (IK)**, **Probability Kriging (PK)**, and **Empirical Bayesian Kriging (EBK)**. Each form of Kriging uses a slightly nuanced formulation and can include a variety of transformations. For brevity, we'll focus on OK for the rest of this section and encourage you to visit `pykrige`'s documentation to learn more about the various types of Kriging available. The documentation can be found at `https://geostat-framework.readthedocs.io/projects/pykrige/en/stable/#kriging-algorithms`.

To implement Kriging-based interpolation in Python, you'll use the `pykrige` package. You'll continue to use the Ohio temperatures dataset that you worked with in the previous section. Follow these steps:

1. To begin, you'll import the required packages:

```
import matplotlib.pyplot as plt
import numpy as np
import pykrige.kriging_tools as kt
from pykrige.ok import OrdinaryKriging
```

2. Identify the minimum and maximum latitude and longitude and create a grid:

```
min_x = min(temps_nogeom['LONGITUDE'])
max_x = max(temps_nogeom['LONGITUDE'])
min_y = min(temps_nogeom['LATITUDE'])
max_y = max(temps_nogeom['LATITUDE'])
# Create the grid
```

```
gridx = np.arange(min_x, max_x, 0.1, dtype='float64')
gridy = np.arange(min_y, max_y, 0.1, dtype='float64')
```

3. Fit the semivariogram:

```
Orid_Krig = OrdinaryKriging(
    temps_array[:, 1], #Longitude vector
    temps_array[:, 0], #Latitude vector
    temps_array[:, 2], #Temperatures vector
    variogram_model="gaussian", #The semivariogram model
    verbose=True, #True writes out the steps as they're
being performed
    enable_plotting=True #True plots the empirical
semivariogram
)
```

Figure 11.7 shows an empirical semivariogram using a Gaussian model. The Gaussian model does reasonably well fitting the points, even with the large outlier:

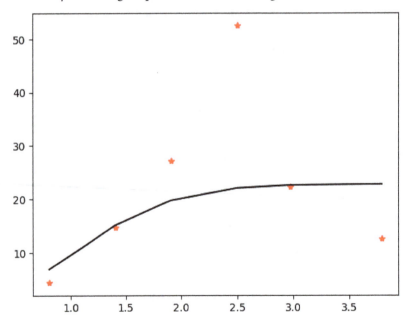

Figure 11.7 – Gaussian-based empirical semivariogram

4. Interpolate the values using the Kriging model:

```
z, ss = Orid_Krig.execute("grid", gridx, gridy)
```

5. Plot the interpolated values and the sampled points:

```
im = plt.imshow(z, extent=[min_x-.05, max_x+.05, min_y-
.05, max_y+.05], origin='lower', cmap='coolwarm')
plt.scatter(temps_array[:, 1],
    temps_array[:, 0],
    c = temps_array[:, 2], alpha=.5, marker='o', s=20,
edgecolors='black', linewidth=1, cmap='coolwarm')
cbar = plt.colorbar(im)
plt.title("Interpolated Temperature")
plt.show()
```

Figure 11.8 shows the interpolated values produced by using OK:

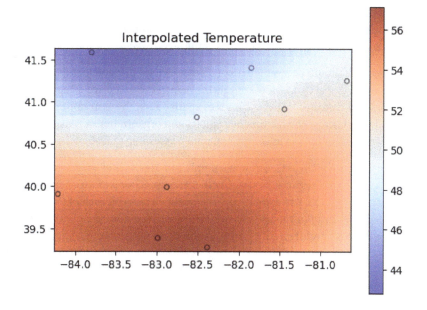

Figure 11.8 – Interpolated average daily temperatures using OK

Compared to IDW, OK produces a surface of interpolated values that is much smoother than that of IDW. The values around the sampled observations are roughly the same as that of the IDW-based approach.

You may now be wondering how you can use these interpolated values in downstream processes. There are many applications where spatial interpolation would be valuable. Pretend that you're trying to forecast energy usage for households served by a local power company. Your original dataset only included temperature information from these nine weather stations. Using spatial interpolation, you can deduce the average temperature across the state and then include the interpolated values as an explanatory variable in some form of regression model to predict energy usage. As you can imagine, there are many other areas where spatial interpolation can be beneficial in your work.

In the next section, we'll transition to a discussion on ethics in spatial data science, which cuts across all steps in your spatial data science workflow.

Ethical spatial data science

When working in data science, GIS, or spatial data science, one thing that you should keep in mind is how to properly use your data in an ethical and responsible way. Interest in the topic of ethics is growing rapidly as companies and organizations must grapple with the meteoric rise of data as one of the most valuable and abundant resources of the modern age. Technological resources are also ever improving our ability to work with large datasets and derive meaning from complex and disparate sources. These technological advantages are not without risks, as the ability to deanonymize sensitive data and the growing number of data hacking incidents where confidential personal details have been made public are something that should keep data practitioners, executives, and leaders up at night.

As a data practitioner, you should be aware that not all data is the same and that data is only as good as the person and process that generates and collects the data in the first place. People tend to assume that data is unbiased and truthful. When it comes to spatial data, the assumption of truth is even stronger as there is a long history of government agencies and organizations releasing maps to guide policy and decision-making, resulting in the general public assuming that all maps are authoritative.

In this section, we're going to briefly walk through a number of examples of when spatial data may not have been collected or used in a responsible way. In other examples, we'll touch on how organizations are proactively trying to prevent harm associated with spatial data. With each of these examples, we hope that it helps spark curiosity in you as to what could have been done differently in this instance and what should be done differently in the future, as the challenges before us are only going to be more complex.

Example 1 – Sharpiegate

In the preface to this book, we briefly spoke about *Sharpiegate* and will take the time now to go into a bit more detail. In September 2019, Hurricane Dorian was barreling toward the east coast of the United States. Then, President Donald Trump presented a doctored NOAA hurricane forecast map with Sharpie lines, indicating that the hurricane would hit Florida and then veer westward toward Alabama. In contrast to the altered map and more in alignment with the NOAA forecast, Dorian moved up the eastern United States, hitting parts of Florida and Georgia. Information stemming from a **Freedom of Information (FOI)** request and then published by *TIME* at `https://time.com/5775953/trump-dorian-alabama-sharpiegate-noaa/` revealed that the map had indeed been doctored after release from NOAA.

Example 2 – Human mobility: The New York Times investigative report

On December 19, 2019, *The New York Times* published an article titled *Twelve Million Phones, One Dataset, Zero Privacy*. The article can be found here: `https://www.nytimes.com/ interactive/2019/12/19/opinion/location-tracking-cell-phone.html`. The article details the curation of human mobility data by cell phone providers and third-party companies. In the article, there are numerous examples where cell phone data was used to create an in-depth history of a person's daily movement. In one such example, it was used to track a senior government official and his wife throughout their day, which included a march in Washington, DC, visits to his office at the Pentagon, interactions with the then President, Barack Obama, and times when he was at home. There are many other incidents like this one detailed in *The New York Times* reporting.

Example 3 – COVID-19 contact tracing

As the COVID-19 pandemic rapidly spread across the globe between 2019 and 2022, governments quickly set up methods for contact tracing in hopes that it would help detect who was exposed to the COVID-19 virus and prevent further spread in the community. Every government contact tracing application had different rules, privacy policies, and options for protecting individuals. Research by Norton Rose Fulbright, published at `https://www.nortonrosefulbright.com/en-cn/ knowledge/publications/d7a9a296/contact-tracing-apps-a-new-world-for- data-privacy`, details the contact tracing applications of various governments and the privacy implications related to those applications. In China, for example, the contact tracing application that was implemented did not require consent from Chinese citizens, and the accessibility, use, and storage of that data was clouded in mystery.

Example 4 – United States Census Bureau disclosure avoidance system

In August 2018, the Census Bureau published an article titled *Protecting the Confidentiality of Americas Statistic's: Adopting Modern Disclosure Avoidance Methods at the Census Bureau*, which is available at `https://www.census.gov/newsroom/blogs/research-matters/2018/08/ protecting_the_confi.html`. The article details the evolution of the Census Bureau's standards for protecting the confidentiality of the data collected from American citizens, households, and businesses. In the early days of the Census Bureau, the only protection that was used was removing the names associated with the collected records. With the 2020 decennial census, the Census Bureau decided to enhance its protections against deanonymization risk by applying **differential privacy**. Differential privacy is based on cryptographic principles and leverages specialized algorithms to ensure that published statistics using an individual's data are no more revealing than statistics where that individual's data was not used. Differential privacy has also been leveraged by Google and other institutions to protect the personal data stored within web browsers and mobile applications. To learn more about the Census Bureau's use of differential privacy, we encourage you to visit `https://`

`www2.census.gov/library/publications/decennial/2020/2020-census-disclosure-avoidance-handbook.pdf`.

In three of these cases, we explored some potentially unethical uses of geospatial data, which caused misleading information to be shared with the public or where the public was tracked largely without its consent. In the fourth example, we discussed how organizations such as the United States Census Bureau and Google are using techniques such as differential privacy to help prevent sensitive data from being disclosed publicly. As an individual practitioner, you may be wondering what you can do to continue to expose yourself to spatial data science ethics. If you were to search the web for data science ethics or codes of conduct, a myriad of frameworks, guidelines, and rules would come up. Each of these frameworks set out with one primary goal in mind: to provide guidance on the best way to responsibly use data. Spatial data science is no exception to this rule as there are a number of guidelines that are published on the web and others that are being discussed and developed.

One resource that we recommend is the letter titled *The Responsible Use of Spatial Data* published in May 2021 by the **Open Geospatial Consortium** (**OGC**). The letter discusses many topics discussed in this section of this book and in many cases goes deeper into areas of concern and areas of hope, including legislation that has recently been passed. It also frames the discussion from multiple points of view, including from users, data providers, and regulators. The OGC acknowledges the shortcomings of its letter, including that it is predominantly written from a Western perspective and is not yet comprehensive in its review of spatial data ethics. The OGC also acknowledges that spatial ethics will change over time, and as such, it is requesting active input and discussion from the geospatial community to continue to improve its documentation and guidance. We hope that this resource continues to be developed over time and that robust standards for the spatial data science community are established.

Summary

In this chapter, we exposed you to a handful of more advanced topics that weren't covered in prior chapters of the book, including spatial indexing and spatial interpolation. Within the section on spatial indexing, we discussed how spatial indexes can be used to improve the efficiencies of spatial queries and spatial operations. In this section, you were introduced to the R-tree index as well as Uber's H3 spatial index, which was used to filter and summarize Airbnb locations in Manhattan. In the section on spatial interpolation, you learned how to infer missing values using sampled points through the application of IDW interpolation and Kriging. In our discussion on Kriging, you were also exposed to variography and semivariograms.

In the last section of this book, we briefly discussed the topic of ethics in spatial data science. We walked you through numerous examples where spatial data has been used in potentially unethical ways, including an altered hurricane forecast, human mobility tracking, and no-consent contact tracing applications used by authoritative regimes. We also discussed how organizations such as the United States Census Bureau and Google are using methods such as differential privacy to ensure that the sensitive data of individuals, households, and businesses are not exposed to the public. The topic of spatial data science ethics is only going to become hotter as technology improves and data resources grow. We're encouraged by the work of the OGC and hope that the spatial data science community works together to create, adopt, and adhere to a set of ethics guidelines to ensure that data is used responsibly.

We hope that you've learned a lot throughout this book and are excited to apply the learnings to your day-to-day work in spatial data science. While our goal was to make this book as comprehensive as possible, we recognize that the spatial data science field is rapidly evolving. As such, we are confident that this book has provided you with a solid foundation from which you can further develop in the future. We also encourage you to give back to the spatial data science community, as there are multiple ways for citizen scientists to contribute to the open geospatial data ecosystem through platforms such as OpenStreetMap. You can also contribute to many of the packages discussed within this book, as several of them are under active development. Lastly, you can contribute to the discussion on spatial data science ethics by visiting the letter published by the OGC.

Index

Packtpub.com

Subscribe to our online digital library for full access to over 7,000 books and videos, as well as industry leading tools to help you plan your personal development and advance your career. For more information, please visit our website.

Why subscribe?

- Spend less time learning and more time coding with practical eBooks and Videos from over 4,000 industry professionals

- Improve your learning with Skill Plans built especially for you

- Get a free eBook or video every month

- Fully searchable for easy access to vital information

- Copy and paste, print, and bookmark content

Did you know that Packt offers eBook versions of every book published, with PDF and ePub files available? You can upgrade to the eBook version at packtpub.com and as a print book customer, you are entitled to a discount on the eBook copy. Get in touch with us at customercare@packtpub.com for more details.

At www.packtpub.com, you can also read a collection of free technical articles, sign up for a range of free newsletters, and receive exclusive discounts and offers on Packt books and eBooks.

Other Books You May Enjoy

If you enjoyed this book, you may be interested in these other books by Packt:

Learning Geospatial Analysis with Python - Third Edition

Joel Lawhead

ISBN: 9781789959277

- Automate geospatial analysis workflows using Python
- Code the simplest possible GIS in just 60 lines of Python
- Create thematic maps with Python tools such as PyShp, OGR, and the Python Imaging Library
- Understand the different formats that geospatial data comes in
- Produce elevation contours using Python tools
- Create flood inundation models
- Apply geospatial analysis to real-time data tracking and storm chasing

Applied Machine Learning Explainability Techniques

Aditya Bhattacharya

ISBN: 9781803246154

- Explore various explanation methods and their evaluation criteria
- Learn model explanation methods for structured and unstructured data
- Apply data-centric XAI for practical problem-solving
- Hands-on exposure to LIME, SHAP, TCAV, DALEX, ALIBI, DiCE, and others
- Discover industrial best practices for explainable ML systems
- Use user-centric XAI to bring AI closer to non-technical end users
- Address open challenges in XAI using the recommended guidelines

Packt is searching for authors like you

If you're interested in becoming an author for Packt, please visit `authors.packtpub.com` and apply today. We have worked with thousands of developers and tech professionals, just like you, to help them share their insight with the global tech community. You can make a general application, apply for a specific hot topic that we are recruiting an author for, or submit your own idea.

Share Your Thoughts

Now you've finished *Applied Geospatial Data Science with Python*, we'd love to hear your thoughts! Scan the QR code below to go straight to the Amazon review page for this book and share your feedback or leave a review on the site that you purchased it from.

https://packt.link/r/1-803-23812-7

Your review is important to us and the tech community and will help us make sure we're delivering excellent quality content.

Download a free PDF copy of this book

Thanks for purchasing this book!

Do you like to read on the go but are unable to carry your print books everywhere? Is your eBook purchase not compatible with the device of your choice?

Don't worry, now with every Packt book you get a DRM-free PDF version of that book at no cost.

Read anywhere, any place, on any device. Search, copy, and paste code from your favorite technical books directly into your application.

The perks don't stop there, you can get exclusive access to discounts, newsletters, and great free content in your inbox daily

Follow these simple steps to get the benefits:

1. Scan the QR code or visit the link below

https://packt.link/free-ebook/9781803238128

2. Submit your proof of purchase
3. That's it! We'll send your free PDF and other benefits to your email directly